Dog Behavior

Dog Behavior
Modern Science and Our Canine Companions

James C. Ha
University of Washington and DrJimHa.com

Tracy L. Campion
Pet Connection Magazine

Academic Press is an imprint of Elsevier
125 London Wall, London EC2Y 5AS, United Kingdom
525 B Street, Suite 1650, San Diego, CA 92101, United States
50 Hampshire Street, 5th Floor, Cambridge, MA 02139, United States
The Boulevard, Langford Lane, Kidlington, Oxford OX5 1GB, United Kingdom

© 2019 Elsevier Inc. All rights reserved.

No part of this publication may be reproduced or transmitted in any form or by any means, electronic or mechanical, including photocopying, recording, or any information storage and retrieval system, without permission in writing from the publisher. Details on how to seek permission, further information about the Publisher's permissions policies and our arrangements with organizations such as the Copyright Clearance Center and the Copyright Licensing Agency, can be found at our website: www.elsevier.com/permissions.

This book and the individual contributions contained in it are protected under copyright by the Publisher (other than as may be noted herein).

Notices
Knowledge and best practice in this field are constantly changing. As new research and experience broaden our understanding, changes in research methods, professional practices, or medical treatment may become necessary.

Practitioners and researchers must always rely on their own experience and knowledge in evaluating and using any information, methods, compounds, or experiments described herein. In using such information or methods they should be mindful of their own safety and the safety of others, including parties for whom they have a professional responsibility.

To the fullest extent of the law, neither the Publisher nor the authors, contributors, or editors, assume any liability for any injury and/or damage to persons or property as a matter of products liability, negligence or otherwise, or from any use or operation of any methods, products, instructions, or ideas contained in the material herein.

Library of Congress Cataloging-in-Publication Data
A catalog record for this book is available from the Library of Congress

British Library Cataloguing-in-Publication Data
A catalogue record for this book is available from the British Library

ISBN: 978-0-12-816498-3

For information on all Academic Press publications
visit our website at https://www.elsevier.com/books-and-journals

Publisher: Charlotte Cockle
Acquisition Editor: Anna Valutkevich
Editorial Project Manager: Ruby Smith
Production Project Manager: Punithavathy Govindaradjane
Cover Designer: Alan Studholme

Typeset by SPi Global, India

Front Cover Image: Ziva, the retired Seattle Police K9. Photograph by Sarah Bous-Leslie.

JCH dedicates this book to his father, Dr. Sam Ha, an enthusiastic ethologist and teacher, and to the mentor who inspired his interest in the world of applied animal behavior, especially dogs: Dr. Philip Lehner.

TLC dedicates this book to her stepdad, Eddie Cesarone, who taught her to love literature and science, and to her dog, Jack, who inspired her to pursue full-time writing.

Contents

Acknowledgments xi
Prologue xiii
Introduction xvii

CHAPTER 1 Dawn of the Dog: Evolutionary Theory and the Origin of Modern Dogs (*Canis familiaris*) 1
    The Birds and the Beaks 1
    Dawn of the Dog 7
    Never Cry Wolf 15
    The Truth About Cats and Dogs 17
    Why Inbreeding is Out 25

CHAPTER 2 Why Tails Wag: Umwelts, Innenwelts, and Canine "Guilt" 33
    Does Your Dog Really Feel "Guilty?" 33
    A Brief History of Ethology 36
    Inside Out 40
    The Matrix is Real 42
    But Wait, There's More 43
    FACS and Fiction 44
    How to Read a Dog 46
    Why Does My Dog Do That? 55
    Found in Translation 58

CHAPTER 3 Mapping the Mind: Brain Structure and Development 63
    Mapping the Mind 63
    If I Only Had a ~~Brain~~ Neocortex 64
    The Not-So-Terrible Twos 68
    How Do Dogs Learn? 71
    Saved By the Bell 72
    The Not-So-Curious Case of Gus Garner 73
    You Talkin' to Me? 75
    Don't Mind the Man Behind the Curtain 77

CHAPTER 4 What A Dog Knows: Analyzing Sensory Perception to Interpret Behavior 79
    The Sixth Sense 79
    Seeing the Forest for the Trees 83

The Nose Knows 85
Heart of Hearing 87
Good Vibrations 88
It's a Matter of Taste 89

CHAPTER 5 The Emotional Animal: Using the Science of Emotions to Interpret Behavior 93
The Age of Denialism 93
Paradigm Shifts 94
The Physiology and Biology of Emotion 98
Making the Case for Canine Depression 100
The Dogs of Ground Zero 103
Arrested Development 103
Why is Emotion so Important? 105

CHAPTER 6 Is It Worth the Risk: How Costs and Benefits Drive Decision-Making and the Evolution of Behavior 109
"Risky" Business? 109
Rules of Thumb 111
Sister Act 112
Hamilton 113
The Secret Life of Pets 115

CHAPTER 7 You Can't Always Get What You Want: The Costs and Benefits of Being Social 121
Time Is On My Side 121
Let Me Be Your Beast of Burden 126
You Can't Always Get What You Want 127
But If You Try Sometimes, You Get What You Need 129
Good Times, Bad Times 134
Gimme Shelter 136

CHAPTER 8 How I Behave Depends Upon How You Behave…Maybe: Game Theory and Our Canine Companions 139
Beautiful Minds 139
The Prisoner's Dilemma 141
To Tweet or Not To Tweet? 145
Equal Pay for Equal Work 147
How I Behave Depends Upon How You're Behaving…Maybe! 148
Canine Brainpower 153

**CHAPTER 9   Debunking Dominance: Canine Social Structure
              and Behavioral Ecology** **155**
    The Socioecology of Social Behavior 155
    Aggression or Dominance? 157
    Of Mice and Men 162
    Breed-Specific Legislation 163
    Debunking Dominance Theory 164
    The Resource Holding Model 170

**CHAPTER 10   The Tale of Our First Friends: Using the Natural
               History of Domesticated Dogs to Understand
               Their Behavior** **171**
    Big Fat Rotten Cows, Dragons, and Piltdown Man 171
    Historia Naturalis 173
    The Lost Boys 176
    From Wolf to Dog 176
    Glances With Wolves 177
    As the World Turnspits 181
    Native American Dogs 181
    Chasing Amy 182
    Dogupations 185
    Dogs and Religion 185
    Pup Culture 186
    Pulling It All Together 188

Conclusion 193
Afterward 195
Index 197

Acknowledgments

James C. Ha
I would like to acknowledge everything I have learned from all of the clients and their dogs: There have been successes, and there have been failures, but every case, every dog, provided an education. I would also like to acknowledge the support of my wife, Renee, who is always willing to listen and improve my ideas. She has contributed a great deal to this book. Finally, I would like to acknowledge the colleagues from whom I have learned so much. Whether through your research and writing, from personal communications with many of you, or via your early reviews of this book (thanks to Patricia McConnell, Lee Drickamer, Celeste Walsen, Ann Howie, and Ann Brudvik), I have learned so much from all of you!

Tracy L. Campion
I would like to thank Dr. Jim Ha for suggesting this collaboration and Dr. Renee Ha for her advice and feedback. I'd like to thank Sarah Bous-Leslie, my business partner with *Pet Connection Magazine* and the amazing photographer for this book; Arianne Taylor and Emilia Galetto, the book's talented artists; Seattle Police Officer Mark Wong and K9 Officer Ziva, who adorns the cover; and Dr. Temple Grandin, Jackson Galaxy, Dr. Patricia McConnell, Dr. Brian Hare, Dr. Emma Goodman Milne, Dr. Barbara J. King, Wendy Spencer of Wolf Haven International, Laurie Hardman of College Dogs, Carly Loyer, and Saethra Fritscher for graciously sharing their knowledge. Thank you to the animals who inspired me and to my mother and grandparents for fostering my love of animals and for providing my formative experiences with dogs, cats, and horses. And last but not least, I'd like to thank my boyfriend, Mark Brad, whose levity and enthusiasm buoyed me throughout this project.

Finally, thank you to the staff at Elsevier, especially the very supportive and efficient Anna Valutkevitch.

Prologue

Alec walked cautiously on the soft underbrush, his dog, Ziva, padding lightly beside him. Her nose traced the familiar trail and she stopped at a tuft of grass, pushing her muzzle into the soil. Finding a warren, she leaned down onto her elbows and burrowed with her nose, tail wagging high above her. Alec called to her impatiently, eager to reach their hunting blind. The early morning sun was just breaking across the far hillside, sending spears of light across their faces. Alec sneezed as the light hit his eyes. Ziva, who was searching the air with her dirt-covered nose, glanced over at him. Alec rubbed his eyes and yawned. Ziva sat down as he paused, yawning as well. Her bushy tail brushed back and forth against the snowy ground, making an arc of naked earth below her. She tipped her nose up and sniffed the air, standing when she caught the scent of something. *Whuff*, she said softly, touching Alec with her nose.

Alec peered toward the brush, his brown eyes adjusting to the waxing light striking the mantled earth. He wasn't accustomed to hiking like this. A bank of new view homes had been put up just above the lake where he'd always hunted, and now he had to travel even farther from the hum of the city – he didn't want to risk shooting near his neighbors' yards.

The foliage closed in on the trail ahead, creating a darkened corridor that opened into a meadow. As they emerged, Ziva spooked at the blind corner, hugging her body against Alec's leg. He patted her and she pushed her nose against him, lingering against his stomach. He'd known her since she was two days old and she'd been his dog since she was two months old. She was almost two years old now, but only recently had she begun her new habit of searching his stomach like this.

Alec continued to walk, signaling Ziva with his hand to stop as he approached the top of the hillside. They were close to his hunting blind, which overlooked a popular watering hole. The wind blew up toward them and Ziva, nose raised, stiffened, her tail going straight up. Alec placed his hand on his hip, ready to fire if necessary. Down in the valley below, a small pack of wolves trotted by. Alec was surprised – wolves here? But perhaps the new housing developments had driven them to this area. Their ragged fur didn't disguise their gaunt frames. A small female wolf toward the back of the pack had a noticeable limp. Ziva stood, transfixed, her own healthy coat catching the growing light.

One of the wolves paused, sniffing the air and scanning the hillside where Ziva and Alec stood, obscured by the trees. Seeing nothing, she trotted on. Game was scarce for everyone. Alec gave Ziva a small piece of jerky – but only one – for remaining quiet. She reached for the bag, wanting more, and stared up at him expectantly, her eyes affixed to his. He gave her one more piece. Ziva walked forward a few steps and then lay down on the snowy grass, looking up at Alec with her jaw slackened. Her expression looked hopeful and expectant to Alec, and he laughed. "No, Ziva," he said, barely audible to his own ears. But her ears flicked back and forth as he spoke. "That's enough. Let's go."

Ziva waited another moment before she saw that Alec wasn't going to stop, and then she trotted on, at first trotting ahead of him, as far as her lead would allow, and then falling back behind him on the narrow trail. She panted quickly, her jaw relaxed. They were getting closer to the blind.

The wind shifted and snow flitted down from a tree branch ahead of them. Ziva sniffed the air in quick, searching movements, her ears flicking forward. Alec watched her closely. The bushes moved almost imperceptibly. Alec squatted down next to Ziva, training his weapon toward the brush. Slowly, a man emerged from the thicket, accompanied by a dog.

Alec stood up slowly, studying the man's unfamiliar appearance. Not wanting to startle the man, Alec waved and called out, "Hello." The man froze, his piercing blue eyes staring up at them, and he cautiously waved back. His dog began to bark, but he pulled it back by the collar sharply, hitting it on the shoulder and saying "Shhh." The dog yipped and sank down, his stomach on the ground. The man gave him a treat from his pocket, which the dog took tentatively. His tail wagged slowly against the ground, but he still cowered.

Ziva leaned forward, her ears flicked back. Her rump rose into the air and her tail slowly wagged back and forth. She let out a quiet whine and then a series of low, breathy pants, like suppressed laughter. Alec patted her reassuringly – Ziva was still young; she wasn't sure if she wanted to protect her owner or play with the new dog. But this new dog was older, and disinterested, and the dog's owner had a tight hold on his lead. Alec and Ziva slowly approached the man, Ziva trotting ahead sideways before she bowed again to the new dog, who did not mirror her invitation to play. Instead, the strange dog curled back his upper lip, exposing his teeth. Ziva dropped to the ground and then onto her back, exposing her stomach. The whites of her eyes showed as the new dog sniffed her over.

"I'm Alec, and this is Ziva. We live just down the hill," Alec began, motioning down the hillside. The man shook his head, smiling nervously. Alec realized that he didn't understand. He gave a closed-mouth smile, nodded, and continued on, shaking his head to himself. Ziva reluctantly followed, pausing and glancing back at the dog. She was one of only three dogs in her neighborhood and opportunities for play were rare.

Alec whistled and gently pulled Ziva's lead. She came back, bumping her nose against his hand, which hung at his side. Alec patted her and smiled, but he was distracted. The influx of foreigners was a growing concern – this man must be a new neighbor, but he hadn't even bothered to learn the local language. He turned back and watched the slight man, who had turned back away from them, his blonde hair blazing in the morning sunlight. His dog trotted along behind him, head low, tail tucked; Ziva watched intently, her own head and tail high. Alec made sure the two were out of sight before he walked the final steps up to his blind.

The blind had a chair for Alec and a soft bed for Ziva. Alec emptied water from his flask into a bowl and Ziva nuzzled his hand before drinking eagerly. He looked down at her face, his eyes dilating as he stared at her. Ziva stared up at him, her brown eyes blackening, as well. He smiled, took a larger chunk of jerky out of his bag, and offered it to her. Ziva lay on her stomach and used a forepaw to hold the dried meat as she chewed on one end. Alec sat in his chair, softly stroking Ziva's brown and silver fur. Finished, she sat up next to Alec and sniffed the air. Alec's hand remained on Ziva's shoulders. Weapon ready, he scanned the valley and watering hole below. Game was scarce for everyone, but he was sure that this time they would find something. Ziva licked his hand and playfully chewed on it with her soft mouth, looking up at Alec. Dropping his hand, she put her head into his lap and sighed contentedly. He smiled and patted her on the head, wondering how many men before him had done the exact same thing.

Introduction

> *"When the Man waked up he said, 'What is Wild Dog doing here?' And the woman said, 'His name is not Wild Dog anymore, but the First Friend, because he will be our friend for always and always and always.'"*
> – Rudyard Kipling

How many thousands of years ago did "Wild Dog" become human's First Friend – and how much have we learned about each other since this occurred? Dog lovers and scientists alike have been debating this for hundreds of years. A mounting body of evidence points to a relationship that began tens of thousands of years ago. In 1994, speleologists (scientists that specialize in studying caves) Christian Hillaire, Eliette Brunel-Deschamps, and Jean-Marie Chauvet had a glimpse of First Friend's early partnership with humans when they were exploring a cave in the Ardeche region of southern France. As they walked through the ancient cavern, they were greeted by a brilliant Paleolithic bestiary adorning the walls – preserved paintings of easily recognizable mammoths, horses, cave hyenas, lions, bears, and even rhinoceroses. But they were also greeted by the accidental remnants of a bygone civilization: Their feet were retracing the steps of prior visitors, many of which likely marveled at – or contributed to – these stony tapestries. Embedded in the cave floor were several sets of prints, including the footprints of a shoeless child, no more than ten years of age and approximately four feet tall, accompanied by the pawprints of a familiar-looking quadruped. The duo had walked together, only feet apart, when the substrate was soft clay. When it hardened, it kept a permanent record of their visit. While the prints looked modern, they were actually created approximately 26,000 years ago. Transecting their 150-foot trackway were other prints, including the tracks of several keystone predators, such as bears and wolves, perhaps contemporaneous with the duo and perhaps made thousands of years before or after their arrival in the cave. The quadrupedal footprints accompanying the child, however, were qualitatively different from the prints of the wild wolves. The pawprint of the animal accompanying the child revealed a four-toed, oval-shaped paw pad with a truncated middle toe, while the wolf prints all had symmetrical toes.[1] The prints, the speleologists observed, closely resembled the prints of modern domesticated dogs, which have shorter middle toes, while wolves lack this derived trait (Fig. 1).

Chauvet Cave, which was later named after Jean-Marie Chauvet, is the site of some of the first physical evidence of the beginning of the human-canine relationship. Scientists were initially hesitant to accept that this partnership predated agrarian societies when the prints were first discovered in 1994, but today, it is generally agreed that dogs and humans were cohabitating long before this child and dog left the impressions that already demonstrate physical differences from those of wild

---

[1] Frydenborg, K. (2017). *A dog in the cave: The wolves who made us human*. New York: Houghton Mifflin Harcourt Publishing Company.

FIG. 1

Chauvet footprints. *Sketch by Arianne Taylor.*

wolves. Yes, prior to 26,000 years ago, there's evidence that dogs and humans were cohabitating, and that dogs as we know them had already begun to demonstrate significant physical and behavioral changes from their wolf cousins. The tracks of dogs differ from those of wolves in both form and locomotory patterns: dogs often drag their toenails along the substrate as they walk, leaving a trackway that's less clear and clean than that of the wolf. Given the age and condition of these prints, it's unclear whether the canids that left these trackways differed in their locomotion, as well.

After hundreds of generations of coevolution, dogs and humans have forged a unique partnership that's unrivaled among animal species. But long before dogs earned their designation as "man's best friend," they were tentatively taking those first steps into our caves, gauging whether this would be a wise alliance, just as much as we were measuring them. Whether canines or humans initiated the relationship – or if it was a confluence of needs among two complementary species – both have been forever changed by the partnership. Tens of thousands of years after the emergence of the human-canine alliance, dogs have become an important part of our everyday lives. Dogs have acquired more than our friendship, though; we've developed a familial bond with them. Two-thirds of those who were surveyed by the American Veterinary Medical Association[2] said that their dogs were family members and almost one-third said that their dogs were friends or pets. Only .7% of respondents

---

[2] http://www.humanesociety.org/issues/pet_overpopulation/facts/pet_ownership_statistics.html.

considered their dogs to be "property" and many who share their lives with dogs self-identify as "pet parents." So how did dogs transition from newfound companion to work partner and from work partner to family member?

From the companions of cave men to police K9s and from herding dogs to service dogs, our canine companions have become an integral part of our lives, and their perceptive skills make them particularly adept at partnering with humans. Dogs can detect cancer cells, low blood sugar, explosives, and illicit drugs; they can predict seizures and alleviate panic attacks; help us herd and hunt; pull our sleds and provide physical support; and they're keenly attuned to our emotional states and even our intentions. Dogs excel at deciphering human communication, often better than humans decipher gestures and other communicative signs among themselves. Dogs can not only read our scents, gestures, and body postures, including facial expressions, they're acoustically multilingual, as well, exhibiting extensive verbal comprehension vocabularies. It's commonly believed that most dogs understand approximately 100 to 200 spoken words, on average. And one linguistic overachiever, a Border Collie named Chaser, reportedly learned a whopping 1,022 proper nouns alone,[3] a vocabulary that rivals that of a human four-year-old.[4] But do we really know dogs as well as they know us?

Over the past 150 years, we have seen dramatic advances in science, from sea and space exploration to medical breakthroughs and from genetics to information technology. We've delved into the Mariana Trench, the deepest point on the planet; discovered that the Milky Way is just one among many galaxies; developed life-saving vaccines; and mapped the earth, stars, and human genome. We've landed on the moon, freckled space with our satellites, and seen the birth and exponential growth of the Internet. We've discovered that our evolutionary next of kin, chimpanzees and bonobos, have complex social lives, use tools, and demonstrate cultural variations that rival our own. We've discovered that interbreeding occurred between *Homo sapiens* and Neanderthals (*Homo neanderthalensis*) and that humans share in common with chimpanzees 98.6% of their DNA. And we've seen dramatic physical changes in dog breeds during this time, too, often without realizing why – or what effects these changes would have on the future of dogs.

Until very recently, relatively little research has been done on the first friend who has been constantly by our side. With all of these scientific breakthroughs, we still know comparatively little about the evolution, behavior, cognition, and communication of those who have coevolved with us for millennia. While humans and chimpanzees last shared a common ancestor approximately eight million years ago, dogs last shared a common ancestor with wolves only 20,000 to 30,000 years ago. Not surprisingly, they share in common with wolves 99.6% of their DNA. Genetically speaking, dogs and wolves are very much alike, but it's that seemingly insignificant .4% that makes all the difference. For decades, that fraction of a percentage hasn't been examined – or it has been mired in flawed research that used captive wolves as a suitable model for domesticated dogs. When a being is said to be domesticated,

---

[3] http://www.chaserthebordercollie.com.
[4] https://www.nlm.nih.gov/medlineplus/ency/article/002015.html.

they are associated with the home and a close association with humans. Granted, the scientific research of our domesticated companion animals has been arguably "less sexy" than discovering and researching exotic species, but a closer examination of canines is as much a story about them as it is about us. And the evolutionary history of domesticated dogs is just as complex and thought-provoking as that of any free-living species.

Just how did domesticated dogs come to be? Behavioral and genetic studies of domestic dogs, as well as the behavior of some free-living wolves, has helped shed some light on this. During the winter of 2003, a lone male wolf in Alaska named Romeo eschewed his wild life, choosing human and dog companionship over that of his fellow wolves. Romeo, who earned his moniker because he was "smitten" with one of the dogs who belonged to a local photographer named Nick Jans, had an assertive but friendly personality. For six years, he stayed near his new friends on the outskirts of Juneau, providing a possible glimpse into the behavioral and personality factors that likely played a part when First Friend approached man, countless times over the centuries. And Russian geneticist Dmitri Belyaev's groundbreaking longitudinal study with silver foxes illustrated how personalities could have been manipulated over dozens, hundreds, and thousands of generations. Belyaev and his research partner, Nina Sorokina, demonstrated how, through successive generations of artificial selection, that fraction of a percentage gradually demarcated protodogs from the ancestors of wolves. Belyaev and Sorokina selected foxes for certain behavioral traits, breeding successive generations of friendly foxes to one another until the species was behaviorally and physically transformed. Despite how seamless the relationship between *Homo sapiens* and *Canis familiaris* appears to be, many research questions still remain unanswered – or even unasked.

We have been asking questions about our four-legged friends since the first protodogs offered their assistance, companionship, comfort, and empathy. Not until relatively recently, though, have domesticated dogs proven to be a rich subject for scientific research. Why and when did dogs and humans first develop their partnership? How can we understand dogs' evolutionary history and the constraints of the species? How and why does the social behavior of our dogs differ from that of our cats? How do our dogs experience the world – and how does their sensory experience differ from our own? Why is it important to understand their brain structure and developmental trajectories? What motivates our dogs to be a part of our families? Why do dogs develop destructive or self-injurious behaviors? Why is "dominance hierarchy" so controversial, and how should we use it – or not? Why do our dogs yawn when we yawn? What does it really mean when our dogs wag their tails? Why do we say that the expression in their eyes "melts our hearts" or is "almost human?" And why do we sometimes feel like they can intuit our thoughts, feelings, and intentions, almost before we realize them ourselves? We wrote this book to help decipher who our canine companions are and *why* they often do what they do – and to make the science behind our canine companions accessible, eye opening, and engaging.

Dogs are unique in the animal kingdom. They're a biocultural construct – the result of both biological variation within the breeding population and the cultural preferences of the humans that chose which dogs would reproduce with one another. As a result, they have long been viewed with the misnomer of artifice, believed to be unworthy of study in comparison to more exotic, free-living species that have not been artificially shaped by humans. But dogs actively participated in their domestication and established a relationship with humans; it wasn't a case of "taming" a wild species, but a great experiment in intra-species socialization that altered the social and evolutionary trajectory of both humans and domesticated dogs. In his book, *How the Dog Became the Dog: From Wolves to Our Best Friends*, American journalist Mark Derr wrote, "We chose [dogs], to be sure, but they chose us, too, and our shared characteristics may well account for our seemingly unshakeable physical intimacy."[5]

Dogs are beloved family members – and we're happy to spend our money on them to keep them happy and healthy. Our dogs are big business, too: they're at the heart of a $60 billion a year industry,[6] with dog-related expenditures estimated at $1,640 per dog annually. Dog owners purchase a wide variety of "accessories" for their dogs, from the standard collars, leashes, bowls, beds, and toys, to food puzzles, screens for safely traveling in the car with the windows rolled down,[7] ramps, seatbelts, and a wide variety of apparel options. According to a recent survey,[8] it's estimated that 54.4 million households have at least one dog – and there are approximately 77.8 million pet dogs in the United States as of 2016. Despite their immense popularity – and their place in the country's economy – there has yet to be a book that provides a comprehensive, accessible guide to canine history, evolution, behavior, and communication.

In a recent tome on the best modern science writing, popular science author Mary Roach wrote, "Make no mistake, good science writing is medicine. It is a cure for ignorance and fallacy. Good science writing peels away the blindness, generates wonder, and brings the open palm to the forehead: 'Oh! Now I get it!'"[9]

This book aims to be good medicine for the human-canine relationship: an antidote to our prior misunderstandings; a pastille for our mistakes; and a prophylactic for future miscommunications. As you learn about our canine companions, we also want you to learn about how modern animal behavior science is conducted. Much of the structure of this book was adopted from an introductory animal behavior course taught to university students who will become the next generation of scientists who will be exploring the ultimate questions of Why and How do we, and they, do what we do. This book aims to present a modern approach to animal behavior science, with a particular focus on our canine companions.

We approach this subject as lifelong dog parents and with the dual perspectives of animal behavior researcher and dog trainer. Dr. Ha has a PhD in zoology (with

---

[5]Derr, M. (2011). *How the dog became the dog: From wolves to our best friends*. New York: The Overlook Press.
[6]https://www.nbcnews.com/business/consumer/americans-will-spend-more-60-billion-their-pets-year-n390181.
[7]www.breezeguards.com.
[8]http://www.humanesociety.org/issues/pet_overpopulation/facts/pet_ownership_statistics.html.
[9]Roach, M. (2011). *The best American science and nature writing*. New York: Houghton Mifflin Harcourt Publishing Company.

an emphasis on behavioral ecology) and has researched animal behavior – and in particular, the behaviors of our companion animals – for more than three decades. During this time, Dr. Ha has also provided in-home consultations for problematic canine and feline behaviors.

One of the reasons that I (JCH) wanted to write this book was to put to rest some of the outdated and just plain odd concepts in "animal behavior" that I heard in the dog behavior world as I became more active in applied ethology during my academic career. I began to hear of concepts, being actively promoted, that were part of the "History of Our Field" lecture in my undergraduate days (which were a LONG time ago, but somehow, seemed to persist). Concepts like Konrad Lorenz's "hydraulic model" of behavior (otherwise known as the toilet bowl model: don't ask, just look it up!) and Fixed Action Patterns (straight from the original work in gulls from the mid-1900s) were fascinating original ideas at the time, but modern animal behavior science has not so much rejected them, but modernized them: the concepts have evolved, morphing into newer, better, more accurate concepts. We need to be applying this newer, modern view of animal behavior science to our canine companions, as well.

"Predatory drift" is another concept that I often hear, but that has no basis in animal behavior science. Introduced by Jean Donaldson at the San Francisco ASPCA in her otherwise top-notch dog behavior training program of the past, this involved something about play or defensive behavior in a few dogs intensifying due to arousal and shifting over into predatory behavior (I apologize to Jean if I am not paraphrasing the concept correctly!). I cannot find any reference to this concept in the animal behavior literature; there's no basis on which to think it would occur in the realm of modern animal behavior science. I understand that it is really short-hand for a series of other processes, but in fact, there is no proven science behind it. Yes, it does make sense…but that doesn't always make good science!

Tracy Campion has Master's degrees in Primate Behavior and Journalism and has researched and written about animal behavior and communication for more than a decade. I (TLC) will provide examples of my own observations with my black Labrador Retriever/Border Collie mix, Jack, a three-legged, one-eyed dog who was rescued from the streets of rural Washington by Seattle Humane after he was hit by a car. I had lived my entire life with a menagerie of animals, including cats, dogs, horses, ponies, a bird, a tortoise, and the occasional rodent rescued from one of my grandparents' barn cats, but Jack was the first dog that I adopted on my own as an adult. I'd specifically been looking for a "differently abled" dog because I'd broken my pelvis and sacrum in 2013 and wanted a dog who could take it slow with me (at first, at least!).

As a primatologist, I've had the opportunity to work with baboons in Africa, spider monkeys in Mexico, and chimpanzees who knew sign language. Jack came into my home one month after his eye and leg were removed and I naturally viewed his behaviors as both a pet parent and an inquisitive scientist. How was he "different" from other dogs I had known – and how was he similar? I quickly saw that he interacted better with people than with dogs, better with puppies than with adult dogs, and that he had a clear affection for horses and goats and a healthy respect (fear?) of the three senior cats who shared our home at the time. Observing him, I noticed things that I'd never before noticed in dogs. This created questions — and it was also the inspiration behind why I became the co-owner of a pet publication called *Pet Connection Magazine*, and I reached out to Dr. Ha about an article. During this first conversation, we discussed the book that he'd been dreaming of writing for a decade, and we decided to take the idea from the intangible and put it into print. So really, I have Jack to thank for all of this.

Having two different perspectives but a shared understanding of animal behavior concepts, we'll explore the origin of modern dogs to understand their genetic history and the constraints of the species. We'll follow this with a discussion of an equally-important fundamental concept: viewing the world through the mind of the animal, referred to as their "umwelt" and "innenwelt." By understanding how our dogs view and process the world, we can better understand what is salient to them and why they behave the way that they do. With these fundamental principles, we can start exploring neuroscience and development to understand the capabilities and limitations of dogs (Chapter 3), the senses and how they influence behavior (Chapter 4), and the far newer and less detailed science of emotions (Chapter 5). All of these systems interact to shape the behavior that we see in our pets and in ourselves. In Chapter 6, we introduce a fundamental approach to the study of behavior itself: the constant struggle to balance costs and benefits. We describe how perceived costs and benefits influence the evolution of social behavior, including the costs and benefits of domestication (Chapter 7), and how the optimal decisions of one individual may depend on the actions of others (Chapter 8). And we will address the complex topic of higher social structure, laden with so much emotional content these days: the questions of dominance, pack theory approaches to training, and aggression (Chapter 9). Finally, we will come back to summarize what we know, and don't know, about our canine companions, and our relationship with them, in Chapter 10.

Throughout these chapters, we'll clarify the mysteries, misunderstandings, and misconceptions about our canine companions with scientific studies, historical evidence, previously unpublished case studies, and our own experiences as animal trainers and animal behavior researchers. We want our readers to look at their dogs and say, "*Now* I know why you do that!" From evolution to social structure and from sensory perspectives to brain structure, we'll map that crucial .4% - and help you understand why dogs truly are humans' first friend.

CHAPTER 1

Dawn of the dog: Evolutionary theory and the origin of modern dogs (*Canis familiaris*)

THE BIRDS AND THE BEAKS

Charles Darwin was fascinated with birds' beaks and dogs' skulls – and for good reason. Avian beaks and canine craniums are morphological markers of evolution, exhibiting natural and artificial selection, respectively, and they provided substantial support for Darwin's groundbreaking theory of evolution – but not necessarily in the way that you may have learned about in school (Fig. 2). "Darwin's finches" have long been referenced as the catalyst behind his theory of evolution by natural

FIG. 2

Charles Darwin.

selection – for decades, these geographically isolated avians have been the biological yardstick of choice. When *The Origin of Species* made its debut in 1859, birds' beaks were referenced dozens of times, yet there was no mention of the famed finches of the Galapagos – and their varied beak shapes and sizes. Darwin did, however, write at length about the vast physical variety seen among domesticated dogs, which should come as no surprise, as he had 13 different dogs of his own throughout his lifetime. From 1831 to 1836, Darwin traveled on the HMS Beagle as a "gentleman naturalist," making observations and collecting specimens as the ship made landfall in numerous countries, including Brazil, Patagonia, the Falkland Islands, the Galapagos, Tahiti, New Zealand, and Australia. In September of 1835, the Beagle made landfall in the Galapagos, and it was the unique geography of these islands and the "finches" that lived on them that garnered the most attention from later scientists, including David Lack's 1947 bestseller, *Darwin's Finches*. Darwin did collect dozens of avian specimens from the islands, including finches, but they weren't the biological epiphany that they've been made out to be. It was the mockingbirds (*Mimus melanotis*) of the archipelago and South America that played a far more important role in his fledgling theory of evolution by natural selection. Darwin collected six South American mockingbirds and four Galapagos ones; he saw differences in size, coloration, and beak length among the island birds that were more significant than the differences that he had seen among all of the mockingbirds of South America.

In his 1839 book *The Voyage of the Beagle*, Darwin wrote: "My attention was first thoroughly aroused, by comparing together the numerous specimens, shot by myself and several other parties on board, of the mocking-thrushes, when, to my astonishment, I discovered that all those from Charles Island belonged to one species (*Mimus trifasciatus*); all from Albemarle Island to *M. parvulus*; and all from James and Chatham Islands (between which two other islands are situated, as connecting links) belonged to *M. melanotis*."[10] Based upon those observations from South America and the Galapagos archipelago, Darwin postulated that birds had been shaped by natural selection, where phenotypic[11] differences influenced differential reproduction and survival of individuals, with variability in heritable traits arising within populations over time. Darwin noted that each of the four Galapagos mockingbird specimens that he collected from four different islands was exclusively found on each island. He didn't know it then, but these 19 islands boast the highest level of endemism on earth. Endemism refers to indigenous species that are unique to geographic locations and cannot be found anywhere else on earth. Of the archipelago's 219 recorded animal species and 600 recorded plant species, 97% of its land mammals and reptiles, 80% of its land birds, and 30% of its plant species are found only in these islands. The Galapagos Islands, which are 600 miles from the nearest landmass, were the perfect place to make observations of isolated species and selective pressures, and thus, evidence of evolution. Perhaps this is why the myth of the Galapagos finches remains such a persistent one. British biologist Julian

---

[10] Page 399.

[11] Observable physical characteristics.

Huxley stated that it was on these islands that Darwin became "fully convinced that species were immutable – in other words, that evolution is a fact." While none of this was included in his landmark tome on evolution, Darwin did write about finch evolution in, *Journal of Researches into the Natural History and Geology of the Countries Visited During the Voyage Round the World of H.M.S. Beagle*, which was first published in 1836. He wrote:

> *Seeing this gradation and diversity of structure in one small, intimately related group of birds, one might really fancy that from an original paucity of birds in this archipelago, one species had been taken and modified for different ends... Unfortunately most of the specimens of the finch tribe were mingled together; but I have strong reasons to suspect that some of the species of the sub-group Geospiza are confined to separate islands. If the different islands have their representatives of Geospiza, it may help to explain the singularly large number of the species of this sub-group in this one small archipelago, and as a probable consequence of their numbers, the perfectly graduated series in the size of their beaks.*[12]

From island to island, the finches and mockingbirds of the Galapagos had noticeable differences with beak size and shape – and in their dietary preferences, as well. Darwin speculated that the ancestors of these birds had arrived on each of the islands many years before, where they were separated by strong winds and tides between the islands and exposed to slightly different environments. These differential selective pressures resulted in birds with certain traits having an advantage – and thus more reproductive success – than others. Over time, four distinct species of mockingbirds and 14 species of finches evolved on the islands, with each differing in its coloration, song, diet, behavior, and beak size and shape. Finches with "parrot-like" beaks specialized in eating buds; birds with grasping beaks specialized in eating insects; birds with probing beaks specialized in eating insects and cactus; and birds with crushing beaks specialized in eating seeds (Fig. 3).

So interested was Darwin with wild birds that after he finished his travels on the Beagle, he took up domestic pigeons, keeping every breed of them that he could obtain. He found the diversity among domesticated birds to be just as astonishing as that of wild ones, given that all of the pigeons had derived from a single wild species, the rock pigeon (*Columba livia*). Darwin wrote: "Compare the English carrier to the short-faced tumbler, and see the wonderful difference in their beaks, entailing corresponding differences in their skulls…" and that the key to their diversity was "man's power of accumulative selection: nature gives successive variations, man adds them up in certain directions useful to him…"[13]

---

[12] Darwin, C. (1845). *Journal of Researches into the Natural History and Geology of the countries visited during the voyage round the world of H.M.S. Beagle*, revised edition (pp. 403–420). Henry Colburn.

[13] Darwin, C. (1859). *The Origin of Species by means of Natural Selection or the Preservation of Favoured Races in the Struggle for Life*. London: John Murray.

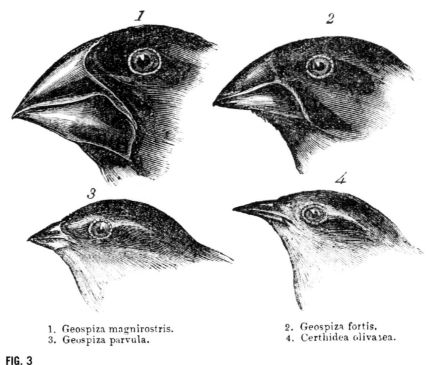

1. Geospiza magnirostris.
2. Geospiza fortis.
3. Geospiza parvula.
4. Certhidea olivasea.

FIG. 3

Sketches of Darwin's birds.

While natural selection was the mechanism behind evolution, Darwin found evidence for artificial selection, as well. The finches and the mockingbirds of the Galapagos had been shaped by natural selection, but the pigeons of Europe had been shaped just as diversely; this time, by humans who had artificially selected them. Humans had artificially selected dogs for generations, as well, intentionally breeding specific individuals with other individuals to preserve or exaggerate certain traits. By favoring novelty and rapidly replicating mutations, humans built dogs with very long (or very short) faces, very short (or very long) legs; dogs with impassive temperaments and dogs who were aggressive; dogs with an excess of wrinkles and dogs with long, silky hair; dogs of every variety. Over time, this resulted in the diverse range of domesticated dog breeds seen today. No other known species, domesticated or free-living, exhibits as much intra-species diversity as the domesticated dog, *Canis lupus familiaris*. Darwin wrote:

> *...domestic [dogs] throughout the world...descended from several wild species, [and] it cannot be doubted that there has been an immense amount of inherited variation; for who will believe that animals resembling the Italian greyhound, the blood-hound, the bull-dog, pug-dog, or Bleinhem spaniel – so unlike all wild*

Canidae – ever existed in a state of nature?... all of our dogs have been produced by the crossing of a few aboriginal species...[and] the possibility of making distinct races by crossing has been greatly exaggerated...[14]

Just as the beaks of the Galapagos finches indicate their dietary preferences, so does the conformation of each dog breed indicate their predilections. With its long, slender legs and deep chest, the Italian Greyhound is a breed built for speed, while the Bloodhound, with its comparatively more robust skeletal structure and long, ground-gracing ears, is built for stamina and can untiringly track a scent for hours. The muscular, stocky Bulldog was originally bred in England during the 1500s for bull baiting, while the diminutive, brachycephalic (short, broad-skulled) pug was initially bred in ancient China to provide companionship for the Chinese ruling class. Nature had produced successive variations of offspring, and rather than being pruned by natural selection in the wild, man had shaped them up in certain directions that were useful to him (Fig. 4).

Both birds and dogs demonstrate the three conceptual cornerstones of Darwin's theory: adaptation, descent with modification, and natural selection. With the influence of humans artificially selecting their mates, dogs in particular show the wide range of possibilities of descent with modification. This observation that newer forms were related to, and descended from, earlier forms, was not new[15]; but Darwin documented the theory of evolution better than anyone who had come before.

FIG. 4

One of Darwin's dog sketches.

---

[14] Darwin, C. (1859). *The Origin of Species by means of Natural Selection or the Preservation of Favoured Races in the Struggle for Life*. London: John Murray.

[15] Based upon his own observations at the Malay Archipelago, naturalist Alfred Russel Wallace delineated an almost identical theory of natural selection that preceded Darwin's. While Wallace and Darwin jointly presented their findings in 1858, it was Darwin who received the lion's share of the credit.

More than 150 years after Darwin's observations, dogs and birds, and every other species, continue to tell the story of evolutionary processes. Wild birds' varied sizes, colors, and beak shapes provide insight into adaptations, descent with modification, and selective pressures. Evolutionary adaptation is defined as the alteration or adjustment in structure or habits which is hereditary, and by which a species or individual improves its ability to survive and pass on its genes in relationship to the environment.[16] Among avians, adaptations include camouflaged feathers, especially with sexually dimorphic[17] game birds; and diversified beaks, such as the ones that Darwin saw in the Galapagos mockingbirds and finches. Among canines, adaptations include paws with claws for traction and fleshy pads to provide better support, as well as thick, camouflaging coats with guard hairs that lock out the moisture, keeping them warm during inclement weather.[18]

Wild birds' adaptations exhibit the effect of selective pressures such as limited or changing habitat, competition, and predation. Birds also provide a considerable body of evidence for the effects of sexual selection. The peacock, with its "costly" tail, demonstrates that individual success can also depend "not on a struggle for existence, but on a struggle between the males for possession of the females; the result is not death to the unsuccessful competitor, but few or no offspring…" Females will only choose partners with these ornamental tails, which marks them as healthy, worthy candidates to father their offspring; thus, individuals "differed in structure, color, or ornament…mainly [from] sexual selection."[19]

Beyond birds and dogs, well-known stories of other species' adaptations illustrate the power of selective pressures. A classic example is a case of industrialized melanism with the peppered moth (*Biston betularia*) in the United Kingdom. At the dawn of the industrial revolution, darker gray peppered moths were relatively rare, but as the revolution progressed over a 45-year span, lichens on the trees died and the trees became stained with soot. The lighter-colored moths, which were once so abundant, were no longer inconspicuous, and predators found them easily. The once-rare darker moths, however, now blended in with the trees, and experienced exponential reproductive success. As pollution emission controls have come into being, however, light-colored lichen has returned to the trees and darker-colored moths have become rare again (Fig. 5). This selection pressure on the moths had a surprising genetic component: The gene involved with these changes, called "cortex," effects cell division and development. A mutation in that gene caused the growth of scales on the moths; with the darker-winged moths having a mutation that caused darker scales. Genetic analyses of canines have revealed surprising results about their physical appearances, as well.

---

[16] The free dictionary.
[17] Distinct size and appearance differences between the sexes of one species.
[18] Compare the coat of the gray wolf to the coat of a Labrador Retriever – one is adapted for warmth and protection, while the other, short, sleek, and providing relatively little warmth or protection, has been selected for by human breeders.
[19] Darwin, C. (1859). *The Origin of Species by means of Natural Selection or the Preservation of Favoured Races in the Struggle for Life*. London: John Murray.

FIG. 5

Industrialized melanism with moths. *Sketch by Arianne Taylor.*

Deoxyribonucleic acid (DNA) is a threadlike double-helix of nucleotides (organic molecules) that's the genetic blueprint for life. DNA doesn't control an organism's behavior, but it specifies the structure of proteins that carry the instructions for an organism's growth, development, and reproduction. A small change in DNA can lead to a dramatic behavioral or phenotypical change, and when this occurs in an amplified way (for example, strong selective pressures via rapid environmental changes or artificial selection), these changes can be both rapid and extreme. The four base nucleotides of DNA are adenine, thymine, cytosine, and guanine, abbreviated as ATCG. Imagine that these nucleotides were represented by four Scrabble tiles (A, T, C, and G) that were placed into a Scrabble bag and shaken up. Each time they were pulled out of the bag, there would be the possibility for all of the letters to change places, providing completely different genetic instructions each time. For example, ATCG might code for floppy ears and a more docile temperament, while TACG might code for erect ears and a more assertive temperament. Thus, just one minor change, much less several major ones, could yield dramatic phenotypic and behavior changes in an organism.

DAWN OF THE DOG

Darwin was rightly fascinated by domesticated dogs' morphological differences, for it is with *Canis familiaris* where we find perhaps the most dramatic example of relatively recent evolutionary history and phenotypic variation. All domesticated dogs descended from a common ancestor of the gray wolf, *Canis lupus*. Modern wolves and domesticated dogs share 99.6% of their DNA in common, but that .4% is a clear demarcation between the species, physically, behaviorally, and developmentally. With its phenotypic variation, *C. familiaris* is unique to the family Canidae, which comprises 34 species, including domesticated dogs, wolves (*Canis lupus*), coyotes (*Canis latrans*), dholes (*Cuon alpinus*), the common fox (*Vulpes vulpes*), raccoon dogs (*Nyctereutes pocyonoides*), and dingoes (*Canis lupus dingo*). Wild canids range in size from the gray wolf, which can reach a shoulder height of 32 inches, to the fennec fox, which reaches only eight inches in height. Of all the species within the family Canidae, none exhibits the range of diversity that is seen among domesticated dogs. From Chihuahuas (which can be even smaller than fennec foxes) to Great Danes (which can stand as high as 44 inches), in their great variety, domesticated dogs tell not only the story of selection pressures placed upon them by humans, but the story of humans, as well, and our desire for companions with very particular traits.

The feliforms (cat-like mammals) and caniforms (dog-like mammals) emerged from the *Carnivoramorpha* clade 43 million years ago, with the *Leptocyon* genus arising 34 million years ago, branching into the Vulpini (foxes) and Canini (canines) branches 11.9 million years ago. The Canids (Family Canidae) first emerged approximately six million years ago on the North American continent, with the first break in the *Canis* lineage occurring one million years ago, with the emergence of the ancestors of modern coyotes. *Canis* is Latin for "dog." The extant *Canis* species include *Canis adustus* (the side-striped jackal), *Canis anthus* (the African golden wolf), *Canis aureus* (the golden jackal), *Canis latrans* (the coyote), *Canis lupus* (including domesticated dogs, *Canis lupus familiaris* and dingoes, *Canis lupus dingo*), *Canis mesomelas* (the black-backed jackal), *Canis rufus* (the red wolf), *Canis simensis* (the Ethiopian wolf), and *Canis lycaon* (the Eastern wolf). Canids' family-based structure was likely the foundation for domesticating them. The process of domestication occurred over several stages, beginning first with taming wild canines, and then selecting them for certain traits, and finally the recent practice of mating dogs to produce separate breeds, which is reflected in the diverse appearance of modern dogs. Despite their morphological diversity, though, all domesticated dogs do belong to the same species. A species can be defined as a breeding population that produces fertile offspring. During intraspecies reproduction, the parents have the same number of chromosomes, so offspring receive two copies of each chromosome, one from the mother and one from the father. The chromosome sets duplicate, and during mitosis, they condense, attaching to spindle fibers and splitting to form new cells. The gametes (sex cells) divide in meiosis, though, taking each set of parent's chromosomes and duplicating and swapping them around before they form four separate cells. During intraspecies reproduction, this is a straightforward process, but during interspecies reproduction, there's a chromosomal mismatch. Individuals from different species that mate and produce offspring (such as a horse and a donkey, with the resultant offspring being a mule, or a lion and a tiger, with the resultant offspring being a liger) often produce infertile hybrids. These hybrids are infertile because there are disparate chromosomes between the parent species, so the offspring don't have viable sex cells. Horses have 64 chromosomes, donkeys have 62, and their offspring, called hinnies or mules, have 63. While *Canis familiaris* and *Canis lupus* are distinct species, they, like all species of the genus *Canis*, can interbreed and produce fertile offspring. While uncommon, wolves and coyotes have been known to reproduce in the wild, producing coywolves, as have domesticated dogs with both wolves and coyotes. Interestingly, the offspring of these interspecies unions are fertile themselves, and evidence of this can be seen in the unfortunate popularity of wolf hybrids as "pets."

According to applied animal behaviorist Dr. Patricia McConnell, breeding dogs to wolves creates "a flood of suffering." She writes: "Wolves simply are not designed to live in houses with people. They need to trot miles and miles every day. They do not, and will never, look to their human for guidance, or boundaries, or anything but to live together as equals. You do not, ever, tell a wolf what to do."[20] McConnell has met a few wolfdogs during her consultation work. She writes: "...they are not wolves, they are not dogs, and they are trapped in the awkwardness of being neither. They can't live in the wild, and many of them can't live with us in our homes, and so they are trapped in a never-never land of never being comfortable in their surroundings..."

I (JCH) have worked on thousands of consultations for animals with problematic behavioral issues, and I have experienced the unique issues with wolf hybrids firsthand, as well. I have largely been based in Seattle for many years, and with the nearness of the Western wilderness, the Cascade and Olympic Mountains, and passion of Seattleites for outdoor life, I have run into a few wolf-hybrid owners. In every case, I found the wolves to be very intelligent, and basically...manipulative. The behavior issues were usually mild, but concerning, aggressive signs (and wolf-hybrid owners generally became concerned about these signs very quickly, too). In each case, I found a dog that was large, bright, and using subtle canine body language to attempt to manipulate the owners into action, whether a walk or exercise (every hybrid that I've ever seen was under-exercised, reflecting the different needs of domesticated dogs versus wolves and wolf hybrids), food, social interaction, or whatever they wanted. There would be a lot of body-blocking and lateral displays, vocalizations (not so much growls, as described by the owners, but "huffs" and "grunts" of displeasure or frustration), even growls and sham-nips or mouthings. As a result, in many cases, owners had rewarded the behavior by reacting in some dog-perspective positive way, and the hybrids were very *quick* to up the ante and continue the game.

The solutions with these hybrids were inevitably: a lot more exercise, to avoid inadvertently rewarding the behavior, and taking your own lead by body-blocking and controlling access to resources, like food. Given their behavioral dissimilarities from dogs, wolf and wolf-hybrid cases have to be taken on a case-by-case basis, with varying degrees of "success" for both the canines and the humans. In these cases, I would recommend that owners implemented this "back-at-you" program very gently,

---

[20] http://www.patriciamcconnell.com/theotherendoftheleash/the-tragedy-of-wolf-dogs.

to see how the dog would react, but in each of the cases I have seen, the wolf-dogs reacted very well, and the relationship was re-configured successfully.

This is not to say that wolves or wolf-dogs are recommended companions for people; almost invariably, they will retain a suite of behaviors that would be appropriate for living in the wild, but maladaptive for sharing a home with humans. But why, even if wolves and wolf-dog hybrids are raised from infancy "like dogs," are they still so behaviorally different from them? It's in their genetic heritage. In an attempt to reproduce the millenias-past act of canine domestication, a 1950s-era Russian geneticist named Dmitri Belyaev and his colleague, Nina Sorokina, bred foxes over multiple generations, finding individual variation among the offspring. Starting with a population of 130 silver foxes (*Vulpes vulpes*), Belyaev observed that some were fearful of humans, while others were not (Fig. 6). He selected for reproduction only those individuals that did not express fearfulness of humans. Over the course of his experiment, succeeding generations not only lost their fearfulness of man, but they lost their pointed ears and puffy tails, as well. In a relatively short amount of time, evolutionarily speaking, a species of canid that had never previously been domesticated came to closely resemble the domesticated dog. Over generations, the foxes' tails became curled and their ears droopy; they began wagging those tails and licking their human comrades. And their color changed, as well: they became black-and-white, a color pattern not seen in wolves, but common in domestic dogs. In 2010, *Scientific American* stated that

FIG. 6

A fox at play. *Sketch by Arianne Taylor.*

Belyaev "May be the man responsible for our understanding of the process by which wolves were domesticated into our canine companions."[21]

When the ancestors of modern dogs were becoming domesticated, they, too, began to exhibit morphological changes from their progenitors. On average, modern domesticated dogs have a smaller brain and head to body size ratio than wolves do, and dogs have a truncated, widened muzzle region in comparison to wolves. Over the generations, the cheek teeth in dogs became crowded and while dogs and wolves have the same dentition patterns, with the adults of each species having 42 permanent teeth, modern dogs' teeth are smaller than the teeth of their wild cousins. Over time, domesticated dogs exhibited a flattening and reduction of the tympanic bullae, changes in the shape of the cranium and mandible, and more rounded eyes. This process is called "neotenization," or the retention of neotenous (juvenile) traits in adult individuals. Recent studies demonstrate that these changes in skull structure likely began to occur shortly after the domestication process commenced.

When early man first selected young, wild wolves for domestication thousands of years ago, at what point during this coevolution did dogs begin to show these behaviors, as Belyaev's foxes did? As we continued to breed select canines for specific traits over many generations, we created dogs who could be docile, or who would track, or hunt, or protect us. But in selecting for desired behavioral traits, we also selected for a large array of other characteristics, just as Belyaev did; facilitating the perpetuation of exaggerated characteristics such as the lithe figure of the Italian Greyhound; the long, drooping ears of the Bloodhound; the grossly enlarged head of the Bulldog; and the short, compressed nose of the Pug. This targeted artificial selection isn't a thing of the past, either; humans continue to try to build a better companion, resulting in purebred dogs with increasingly exaggerated traits and with creating "designer dogs" by mixing two or more breeds of "purebred" dogs. Examples of recently "designed" dogs include the Labradoodle (Labrador Retriever/Poodle mix), Goldendoodle (Golden Retriever/Poodle mix), Cockapoo (Cocker Spaniel/Poodle mix), and Pugle (Pug/Beagle mix).

Mixed breed dogs with poodle heritage are often marketed as being "hypoallergenic" by their breeders and carry a hefty price tag of $2,500 or more because they "shed less" than dogs who don't have poodle pedigrees. Unfortunately, this isn't accurate: Allergies don't originate in the hair, nor are they dependent upon the type of coat that a dog has; they originate from an animal's dander. A 2011 study in the *American Journal of Rhinology & Allergy* found that across 60 dog breeds, 11 of which were classified as "hypoallergenic," there was no statistically significant difference in the amount of allergens that were present in the home. The prime dog allergen, Can f 1, was present at equal rates across all of the homes in the study[22]

---

[21] http://blogs.scientificamerican.com/guest-blog/mans-new-best-friend-a-forgotten-russian-experiment-in-fox-domestication.

[22] Nicholas, C. E., Wegienka, G. R., Havstad, S. L., Zoratti, E. M., Ownby, D. R., & Christine, C. J. (2011). Dog allergen levels in homes with hypoallergenic compared with nonhypoallergenic dogs. *American Journal of Rhinology and Allergy, 25*(4), 252.

providing no environmental evidence for the hypoallergenic claim. Despite this, articles expounding the virtues of these designer dogs continue to abound.

While they have a shared common ancestor, modern wolves and domesticated dogs have divergent evolutionary trajectories. Although they can still produce viable offspring together, wolves differ from dogs genetically, socially, behaviorally, neurologically, and developmentally. How might the ancestors of modern dogs and wolves have lived? Archaeological sites dating as far back as the Middle Pleistocene period (781 to 126,000 years ago) have bones from both early hominins and early wolves, but in what capacity were they existing? What motivation would wolves have had to go with man, all of those thousands of years ago? How would the benefits of domestication have outweighed the costs? Would Ziva's ancestors have chosen to go with man for controlled but reliable resources, for protection and companionship, instead of trying to hunt for limited resources, just like the wolves that Alec and Ziva saw? Did wolves select humans who would be suitable for them, just as humans selected wolves? In Juneau, Alaska, in 2003, a young male wolf did just that. After a pregnant female wolf (likely his mate) was struck and killed by a taxi, the lonely male took an interest in a wildlife photographer and his dogs. The handsome, sooty-coated wolf was so enamored with one of the photographer's female dogs, that he earned the moniker "Romeo." Romeo, who lived in Alaska's Mendenhall Valley, was an Alexander Archipelago wolf (*Canis lupus ligoni*), a rare gray wolf subspecies endemic to the islands of Southeast Alaska. The remaining Alexander Archipelago wolves were likely remnants of a larger population that had once inhabited lower North America during the last glacial period. Fewer than 1,000 of these wolves exist today, which made Romeo a particularly important member of a keystone species.

Rather than seeking out the companionship of other wolves, Romeo lived near people for years; over time, he became bolder, actively seeking out people and engaging their dogs in play. He most enjoyed spending time with his "favorite" dog, a Labrador named Dakotah, and he was a popular attraction for locals and tourists alike. When Romeo disappeared in 2009, his absence was felt profoundly, not only as the loss of a beloved individual, but also because he provided a window into how the protodomesticated ancestors of modern dogs may have chosen to join the ancestors of man. The wolf who had tried so hard to befriend man had a demise as tragic as his namesake: His body was found that September; two hunters had illegally gunned him down. And while Romeo's attempt to join man was ultimately unsuccessful, it provided an illustration of just how easy – and just how hard – it must have been, thousands of years ago, for wolf to trust man, just as much as it was for man to trust wolf.

Wolves and, later, protodogs were justified in their hesitancy to trust humans. A study published in the *Journal of Archaeological Science: Reports* provides evidence that early on in the human-canine relationship, dogs oscillated wildly between being a convenient food source and being a pet or guard animal.[23] While some

---

[23] Ewersen, J., Ziegler, S., Ramminger, B., & Schmolcke, U. (2018). Stable isotopic ratios from Mesolithic and Neolithic canids as an indicator of human economic and ritual activity. *Journal of Archaeological Science: Reports, 17*, 356–357.

dogs, like the Chow-Chow, have long been bred to be a food source, twenty-first century dog consumption is less common. Some cultures still consume dog meat, though: Dogs are on the menu in Korea, China, Vietnam, Indonesia, the Philippines, Polynesia, Taiwan, Ghana, Liberia, Cameroon, the Arctic, and the Antarctic, and in some parts of Switzerland. The researchers in this study surveyed prehistoric dog meat consumption using dog bones from the Mesolithic and Neolithic periods. They analyzed cut marks and isotopic ratios in the collagen of the dog remains and compared their findings to wolf remains from a comparable time period and geographic area. Compared to the wolf bones, the dog bones had higher rates of carbon and nitrogen isotopes, which would mean that the dogs were eating more vegetation and seafood than the wolves, whose isotopes revealed a carnivorous diet. These isotopic differences reflected the dogs' change from carnivorous hunters and scavengers to opportunistic omnivores. Consuming dog meat is a "taboo" in most industrialized countries, but this early "blurring of the lines" is similar to the relationship that Bedouins have with their Arabian horses. This is perhaps the closest human-nonhuman animal partnership outside of the one established with dogs and people. When Bedouins and their horses are out in the desert and low on provisions, a Bedouin will cut a vein on his horse and drink its blood. This might be an alarming prospect, but consider this: blood is high in iron and in situations where dehydration is a mortal risk, it can provide temporary hydration and fuel.

There are several hypotheses pertaining to how dogs became domesticated, including the "Garbage Dump Hypothesis" and the "Pet Hypothesis," wherein wolf pups were tamed and used for hunting and learned to become part of a new family comprised of canines and humans. Dr. Brian Hare, director of the Duke Canine Cognition Center, believes that man did not domesticate wolves, but instead, a scenario similar to Romeo's occurred multiple times over the years. Dr. Hare finds fault in the hypothesis that man tamed protodogs to help him hunt, stating that man was already an accomplished hunter by the time that wolves began to share his hearth; it was wolves who found the company of man advantageous, many thousands of years ago.

"I think that dogs evolved after wolves chose us as we started to become more settled and move around less as a species," Hare explained. "Wolves just took advantage of the garbage that we created, which is exactly what they eat, and over time, the wolves that were friendly evolved into dogs." This "garbage dump hypothesis" has wide-ranging support, grounded not only in modern domesticated dogs' propensity to raid the garbage can when given the opportunity, but also with the presence of efficient starch metabolizing genes. Dogs and wolves both have these genes, which are needed to process starch, but dogs use them far more efficiently than their wild cousins do.[24]

While humans today have bred dozens of dog breeds, Hare doesn't believe that the species' origin came at the hands of people. "I don't think that people [initially created dogs] – the friendly wolves were actually at an advantage over the wolves that were afraid of people," he said. "If you were running away from people, you couldn't

---

[24] https://www.bbc.com/news/science-environment-21142870.

take advantage of the garbage that they were creating, but if you were friendly and excited to be near people, then you could take advantage of that new, reliable food source. Just like the Belyaev foxes happened artificially through artificial selection, wolves evolved into dogs through natural selection." And, just like Belyaev's foxes, they did so relatively quickly.

The descent of domesticated dogs from the ancestors of gray wolves is widely supported by the scientific literature. In a 2003 paper entitled "Population Genetics, the dog that came in from the cold," researchers Acland and Ostrander wrote: "The dog, beloved as humankind's most faithful companion, descends from the wild gray wolf, but how this happened has been a matter of conjecture, controversy, and confusion."[25] A 2015 study[26] that drafted the genome sequence of a 35,000-year-old wolf from northern Siberia provided evidence that the ancestry of modern-day dogs can be attributed to multiple wolf populations and suggested that the direct ancestors of modern dogs diverged from wolves prior to the Last Glacial Maximum 20,000 years ago.

Early on in the domestication process, dogs benefited from a perfect storm of fecundity (fertility and the ability to produce offspring), opportunity (living in close proximity to other dogs to breed to), and an impressive lack of discrimination. Dogs, unlike wolves, aren't picky about who they'll mate with – they're indiscriminate and often very ambitious breeders. Male wolves who reproduce are typically paired with one female for their reproductive lifetimes (unless she dies, at which time they might find a new mate), but male dogs are very promiscuous. Interestingly, there's more genetic variation in the maternal DNA (mtDNA) of dogs than there is with paternal DNA, indicating that certain male dogs were favored for breeding – a practice that continues today with pedigreed dogs and horses. Examining the mtDNA of modern dogs reveals that there are both recent and "ancient" dog breeds – animals that have been artificially selected by humans for hundreds of years. These breeds include the Asian Akita, Chow-Chow, Shar-pei, and Shiba Inu; the Arctic Husky and Malamute; the Middle Easter Afghan Hound and Saluki; and the African Basenji.

With rare exception, behavior doesn't fossilize,[27] so it's hard to say exactly what drove those early canids toward alliances with early man. All of those animals are now extinct, of course, from the Pliocene-era (23-5.3 million years ago) precursors of the wolf to the most recent shared ancestor of dogs and modern wolves. In 1997, researchers conducted the first thorough examination of dog maternal DNA (mtDNA) on multiple canids, including wolves, coyotes, jackals, and dogs. There are several "ancient" types of dog breeds, including the Australian dingo, the Balinese

---

[25] Acland, G. M., & Ostrander, E. A. (2002). Population genetics: The dog that came in from the cold. *Heredity, 90*, 201–202.

[26] Skoglund, P., Ersmark, E., Palkopoulou, E., & Dalen, L. (2015). Ancient wolf genome reveals an early divergence of domestic dog ancestors and admixture into high-latitude breeds. *Current Biology, 25*, 1515–1519.

[27] Rare exceptions include the 26,000-year-old human and canine footprints in Chauvet Cave and fossils of 67 million-year-old snakes eating baby dinosaurs – effectively, fossils preserving behavioral moments in time.

kintamani, the Asian Akita, Chow-Chow, Shar-pei, and Shiba Inu, the Arctic Husky and Malamute, the Middle Eastern Afghan Hound, and the New Guinea singing dog.

It isn't just "when" and "how" dogs came in from the cold that has been under increased scrutiny; it's also the why and "if" dogs descended from the common ancestors of gray wolves. Did dogs descend from the common ancestor of gray wolves tens of thousands of years ago? And if not gray wolves, then who? While Belyaev's work demonstrated that the process of domesticating dogs could have occurred during a relatively short amount of time, perhaps it also demonstrated that dogs didn't come from the ancestral canids we've long suspected.

Researcher Janice Koler-Matznick, who has worked with the aboriginal New Guinea singing dogs for more than one quarter of a century, believes that this just might be the case. In her 2016 book *Dawn of the Dog, The Genesis of a Natural Species*, Koler-Matznick presented the Natural Species Hypothesis that dogs are wolves, but they aren't derived from *Canis lupus*. She writes: "…the dog is a distinctive biological species of wolf, an exceptional one with unique traits compared to other wolves."[28] Koler-Matznick believes that dog-origin hypotheses such as Purposeful Domestication (the "Pet Hypothesis") and Self-Domestication (the "Garbage Dump Hypothesis") are fallacies that fail to completely explain how dogs as we know them came to be a part of the human household. An historical overview of the human-wolf relationship illustrates just how contentious the question of dog origins remains.

NEVER CRY WOLF

There is perhaps no animal that is more culturally divisive than the wolf, who is simultaneously majestic and menacing, emblematic and problematic. The gray wolf is physically formidable: it's the largest of the free-living canids, standing 30 inches at the shoulder, armed with long canine teeth, weighing as much as 145 pounds, and clad in a coat varying from the typical gray coloration to black, white, reddish, brown, yellowish, or variations thereof. Wolves graced the cave walls in Font-de-Gaume, southwestern France – works of art that were discovered in 1901 and created some 17 to 20,000 years ago – evidence of their importance to early humans. The wolf depiction was both carved and painted into the rock, with only the carved portion of the artwork surviving to modern day (Fig. 7).

Wolves have been hunted throughout human history, though; while they were once common throughout Europe, by 1559, Thomas Platter wrote that the Tower of London's "lean, ugly wolf" was the last one of his kind in England. According to some, the last wolf of England died in 16th Century England by Henry VII's order. Prior to the arrival of European immigrants, as many as 500,000 wolves resided within the continental United States, but over the years, these numbers had dwindled

---

[28] Koler-Matznick, J. (2016). *Dawn of the dog: Genesis of a natural species.* Central Point, Oregon: Cynology Press.

FIG. 7

Prehistoric dog sketch.

drastically. Even environmentalist Theodore Roosevelt called for their eradication. By 1960, only 300 wild gray wolves remained in the continental U.S., and the population has slowly rebounded, thanks to long-term conservation efforts. As the range of wild wolves in the U.S. has slowly spread once again, wolf advocates have heralded their return to areas that they historically inhabited. Yet similarly sanctioned hunts of wolves and coyotes continue in the United States and Canada today: these apex predators are often culled from helicopter, with bounties placed on their heads. So how did the ancestors of today's dogs move from vilified to beloved – and then to the foundation of today's $60 billion pet industry? Dr. Hare believes that wolves who were bold, but friendly – much like Romeo was – approached humans. These wolves were tolerated, and over the succeeding generations, tolerance grew into companionship – and mutual understanding. According to Hare, "These protodogs evolved the ability to read human gestures." An incomparable partnership was born.

Despite Dr. Hare's disagreement with the hunting hypothesis, modern day hunters who use dogs have significantly more success than those who hunt without dogs. Many attempts have been made to ban the use of dogs with modern day hunting due to claims that dogs provide an "unfair advantage." Numerous studies have supported the efficacy of hunting with dogs. In Finland, moose hunters who were accompanied by a dog obtained 56% more prey than those who hunted without one.[29] Hunting with dogs yielded more success for tribes in Nicaragua[30] and Brazil,[31] as well. Another study noted that if hunting dogs helped carry the meat, their human companions would be expending less energy; thus, each kill could provide a greater net gain.[32]

It's highly likely that dogs became protectors, hunting partners, and companions early on during the first instances of canine domestication. In addition to providing

---

[29] http://www.sekj.org/PDF/anz41-free/anz41-545.pdf.
[30] https://www.jstor.org/stable/10.1086/592021?seq=1#page_scan_tab_contents.
[31] http://ethnobiomed.biomedcentral.com/articles/10.1186/1746-4269-5-12.
[32] https://www.americanscientist.org/libraries/documents/2012040911041041krbcmt8m/2012491121109042-2012-05MargShipman.pdf.

help with hunting, dogs were protectors, offering early warning signs for intruders, just as Ziva did on their walk to the hunting blind. The ancestors of Romeo and Ziva alike underwent changes that were later reflected in their descendants – changes such as Ziva's friendly temperament and Romeo's darker, shorter, coarser hair and smaller body size in comparison to wolves that resided in the Alaskan interior. According to Darwin, this is descent with modification, and it occurs when later species are derived from earlier species following changes in structure or behavior (adaptations) required for survival. In wild environments, descent with modification afforded certain animals competitive advantages in comparison to their conspecifics, which resulted in increased reproductive success for some individuals and their families. Darwin wrote that divergence occurred when two evolving groups of recent common ancestry gradually became more dissimilar. Canines have experienced numerous divergences in their evolutionary history, with the first documented break within the genus *Canis* producing what would become the coyote one million years ago. According to a 2015 study,[33] the ancient wolf genome reveals that dog ancestors diverged from the ancestors of modern wolves at least 27,000 years ago. Numerous studies estimate that dogs diverged from the ancestors of European wolves as long as 32,000 years ago. Irrespective of when the genetic lineage diverged, domestication appeared to be a strong selective pressure long before this divergence occurred. Scientists studying dog and wolf DNA found that dog domestication could have first occurred 100,000 years ago or even earlier.[34]

As the ancestors of dogs and wolves diverged, the ancestors of modern man and modern domesticated dogs converged. Conversely, convergence occurs when two evolving groups of distant common ancestry become more similar. While the classic example of convergent evolution is the development of wings between both bats and birds, some have argued that humans and canines are coevolving and converging, as well.[35] In domesticating dogs, humans changed the directionality of their own lineages, as well. Some studies have hypothesized that humans drove out their Neanderthal rivals (and sometime mates) after taming and breeding wolves to help them hunt in Europe, tens of thousands of years ago.[36]

THE TRUTH ABOUT CATS AND DOGS

But what about cats? Along with dogs, cats are one of the only domesticated animals that have lived within our homes for generations, but far less is known about our feline family members than our canines. While dogs have co-evolved with humans for 32,000 years or longer, cats came to share their lives with humans comparatively

---

[33] https://www.cell.com/current-biology/abstract/S0960-9822(15)00432-7.
[34] Vilà, C., Savolainen, P., Maldonado, J. E., Amorim, I. R., Rice, J. E., Honeycutt, R. L., et al. (1997). Multiple and ancient origins of the domestic dog. *Science*, 276(5319), 1687–1689.
[35] http://animalwise.org/2011/09/26/converging-with-canines-are-humans-and-dogs-evolving-together.
[36] http://news.nationalgeographic.com/news/2015/03/150304-neanderthal-shipman-predmosti-wolf-dog-lionfish-jagger-pogo-ngbooktalk.

FIG. 8

Humans chose to bring cats with them as they settled into an agrarian lifestyle. *Sketch by Arianne Taylor.*

recently. According to some studies,[37] the advent of cat domestication was approximately 9,000 to 12,000 years ago, and the first known instance of a cat buried with a human happened 9,500 years ago in a shallow grave in Cyprus.[38] Cats aren't native to this area, so the grave provides evidence that humans were choosing to bring cats with them as they settled into an agrarian lifestyle (Fig. 8).

It is far more rare to find feline remains than canine ones at ancient archaeological sites, in part due to the fact that dogs were sometimes a convenient food source. It appears that the ancestors of modern man didn't consume cats as much as dogs, however. The timeline of feline domestication, unlike that of domesticated dogs, coincides with the rise of agriculture. Farmers needed help fending off rodents from their grain and with easy access to prey, comfortable housing, and the occasional handout, cats were happy to assist.

Cats, like wolves, have had a turbulent relationship with humans, being treated either as near-Gods or as signs of the devil. Cats in several cultures, including ancient Egypt, Islamic countries, and ancient China, were revered. For ancient Egyptians, cats were given preferential treatment over other animals and were associated with the goddesses Ba'at, Bastet, and Isis. It was illegal to kill a cat and they would be embalmed

---

[37] http://www.pnas.org/content/111/48/17230.abstract.
[38] http://news.nationalgeographic.com/news/2004/04/0408_040408_oldestpetcat.html.

and mummified after their deaths and would be mourned by the entire family. In China during the Song Dynasty, cats were highly esteemed as rodent catchers and companions. Muhammed reportedly had a favorite cat named Muezza, and Muhammed, like the ancient Egyptians, placed prohibitions on killing cats. The Vikings kept cats both as companions and as vermin catchers. Ancient Greeks and Romans associated cats with their goddesses Artemis and Diana. Cats appeared to be just as mysterious to the sages of old as they are to modern experts. Greek and Roman philosophers and writers associated cats with cunning (Aesop), lechery (Aristotle), and cleanliness (Plutarch.)

For centuries, felines had been adored at best and viewed as utilitarian companions at worst, but then the tide turned for them. During the Middle Ages, cats were considered to be the companions of witches, a strange belief that resulted in their avoidance or deaths and persisted throughout the Renaissance. During the 14th century, as the Black Death swept across the European continent, cats were seen as carriers of plague and they were killed in droves. While the medieval Welsh King Hywel Dda made it illegal to harm or kill cats, cultural viewpoints of cats were already altered. For centuries afterward, cats were perceived to be harbingers of bad luck, exemplified by the adage, "don't let a black cat cross your path."

With 12,000 years or less of domestication, though, the house cat (*Felis catus*) genome still shares much in common with its wild relative (*Felis silvestris*).[39] In terms of genetic differentiation, cats and dogs have similar differences from their wild ancestors to the modern-day animals who share our homes. Domesticated cats may be separated from their wild descendants by 13 genes,[40] while domesticated dogs and wolves are separated by 11 fixed genes.[41] Compared to the 400 known dog breeds, though, the Cat Fanciers' Association currently recognizes only 41 cat breeds, varying from the diminutive, four-pound Singapura to the hairless, medium-sized Sphynx to the longhaired Norwegian Forest Cat, which often tips the scales at 16 pounds or more. And each of these is "less domesticated" than a dog, say researchers at the Washington University School of Medicine. Their study states that cats are only "semi-domesticated," having split off from their wild ancestors so recently. The study suggests that, similar to dogs, "…selection for docility, as a result of becoming accustomed to humans for food rewards, was most likely the major force that altered the first domesticated cat genomes." There's a reason why people describe their cats as "gracing them with their presence" – one study states that cats are "only domestic when they want to be."[42] The differential selection pressures that were placed on cats versus dogs can at least partially explain this. Humans selected dogs who would have been good at hunting, guarding, or being companion animals, while cats would not have been selected for any of these purposes. From the onset, cats fit a different cultural

---

[39] http://www.smithsonianmag.com/smithsonian-institution/ask-smithsonian-are-cats-domesticated-180955111/#WRuWUXDELyCVC1tI.99.

[40] http://www.latimes.com/science/space/la-sci-sn-cat-genome-20141107-story.html.

[41] Cagan, A., & Torsten, B. (2016). Identification of genomic variants putatively targeted by selection during dog domestication. *BMC Evolutionary Biology*, *16*, 10.

[42] https://www.smithsonianmag.com/smithsonian-institution/ask-smithsonian-are-cats-domesticated-180955111.

niche with humans than dogs have. As vermin exterminators, they were part of a household, and yet removed; they have never been quite as understood as dogs have been. Cats were revered as Gods in ancient Egypt, reviled as witches' companions during the middle ages, and are today the most popular house pet in the U.S., numerically speaking: there are more than 90 million domesticated cats residing within 34% of U.S. homes.[43] Thus, the artificial selection of dogs likely began fairly early on during the human-canine relationship, while there is a shorter known history of selectively breeding cats (Fig. 9).

While they share their lives with us, dogs and cats are very different from one another, genetically and socially. Cats are obligate carnivores, meaning that they evolved to almost exclusively eat meat. Dogs, on the other hand, are opportunistic omnivores that can digest a wide variety of foods; some of them even thrive on vegan diets. Cats come from a long line of solitary hunters and they are relatively self-sufficient. They have co-evolved with humans for roughly one quarter of the time that dogs have. Dogs like Ziva, however, are pack animals that have coevolved next to their people for millennia, descending from a long line of animals that had been artificially selected by humans for companionship, protection, and service. In selecting for these traits, humans molded the directionality of canine evolution, yielding results as divergent as the teacup Chihuahua and the massive Great Dane – animals whose morphological characteristics provide the outward appearance of different species, but whose chemical makeups reveal a shared ancestry not from several species, as Darwin had proposed, but from one. And with practical impediments to procreation from such divergent

FIG. 9

Cats have been artificially selected by humans for less than 10,000 years, a fairly short time in comparison to dogs. *Photograph by Sarah Bous-Leslie.*

---

[43] https://www.smithsonianmag.com/history/a-brief-history-of-house-cats-158390681.

members of one species, speciation is likely in the not-so-distant future for canines, or perhaps has already occurred, in the case of some of the most highly dissimilar breeds.

University of Washington student Saethra Fritscher has a Master's of Science in animal behavior and is currently in a PhD program focusing on applied animal behavior – and feline behavior, in particular. After working in the veterinary field for more than 15 years, she observed that many of the issues were behavioral, and not physical, and that animals with no health issues were being euthanized. She focused her research on felines to help diminish euthanasia rates and educate pet owners on providing the best possible care for their animals.

"The biggest difference when comparing cat and dog behavior is their social styles," she explained. "A dog's social style is so much more similar to our own. Cats are certainly social, but their social behavior seems so different from ours and from dogs, that owners miss a lot of social information!"

Saethra has worked on cases where both dogs and cats lived in the household, but she has never specifically had to address a cat-dog conflict. "If I had to guess, I would attribute this to the fact that owners are pretty intolerant of dogs aggressing towards cats (as they should be, due to the potentially life-threatening repercussions), and may rehome one or the other if such behavior persists," she said. "And when cats aggress towards dogs, dogs quickly learn to leave them be!"

It's important to address all health issues that a cat must have before addressing any behavioral issues.

"I can't stress enough how important it is to adequately test for and treat underlying physiological problems in cat behavior cases!" Saethra explained. "The vast majority of feline inappropriate urination has at least some physiological component, and a good portion of cat aggression does too. Behaviorists should work closely with veterinarians, because if the underlying physical issue is not resolved, the behavioral modification is not going to work."

I've never been called in to help integrate cats and dogs within a household. But, in general terms, cats do best with very slow, positive, and calm introductions. Start the relationship off on the right foot by introducing animals much more slowly than you think you need to, because it's very hard to recover from a negative first impression!

While cats continue to be mysterious for a lot of pet owners, understanding their behaviors can improve your relationship with your feline.

Your cat is not plotting to kill you and your cat loves you just as much as your dog does!

Dogs are like little kids, while cats are like grown-up friends – both are valuable relationships, just different. One species is not better than the other!

In recent years, Jackson Galaxy, also known as the "Cat Daddy," has achieved acclaim for approaching feline advocacy from a feline's perspective. Jackson, who's a two-time New York Times best-selling author and host and executive director of the Animal Planet program *My Cat From Hell*, has worked as a cat behavior and wellness consultant for more than two decades.

"The more that you think about cats as another species, the more distant you are from them," he said. "Working with any species, whether it's human or animal, there has to be room for both science and empathy. My clients realize that it's fifty-fifty between them and their cats; it's a relationship, and all relationships function under certain rules. Just because one has four legs and one has two and one has more power, that doesn't mean that it's not still a relationship."

While our relationships with our cats can suffer because we struggle to read our cats' behavioral cues, how well do our dogs respond to them, and how can these responses be used to assess compatibility with cats in an adoptive home? According to a study published in *Applied Animal Behaviour Science*,[44] shelter dogs were more responsive to auditory than visual or olfactory cat-related stimuli. This provides support that cat sounds could predict which shelter dogs would probably do well when placed in a home with cats or other small animals and it also provides evidence of interspecies differences in communication and the practical application of animal behavior principles.

When Jack, then 18 months old, was adopted into a home that had three senior cats, he had a steep learning curve. Jack had been housed in a foster home with two cats, and his foster parents had reported that he was "appropriately" afraid of them. In his new home, over time, he learned how to cohabit with them peacefully, watching them closely for indications of what they might do, just as closely as they were watching him. During the first six months, meows and hisses elicited startle responses from him (he would yelp, or jump, or run, or some combination of the three), but after one year, both species appeared to understand and accept one another. It would be at least another year before something resembling "friendship" began to bud between Jack and the oldest cat, Meg, who was approximately 20 years old at the time.

Not all dogs and cats can live with one another; some dogs have a high prey drive, and want to chase any smaller animals, and some cats remain fearful and reactive around dogs. Cats and dogs also communicate in fundamentally different ways, including the use of certain vocalizations, tail position, ear placement and movement, and body posture. Mistaking these signals can lead one of the animals to believe that the other animal is aggressive or receptive when they aren't. Dogs and cats do seem to find some common ground with certain behaviors, such as growls and yelps, which appear to have a universally negative connotation, grooming (if they groom one another, this is referred to as "allogrooming"), tension around the mouth, which denotes aggression in cats, dogs, and humans, as well, eye contact and blinking ("soft," frequent blinks indicate affable interactions), and resting, where both species will share space with each other while they are sleeping or relaxing. It was through allogrooming (senior cat Meg, who was 20, would lick Jack's face and ears) and resting, with and without contact, that Jack and the cats began to cement their relationship.

Unlike cats, who have only been selectively bred for a few centuries, dogs have been artificially selected for thousands of years. So when did inbreeding and line breeding of closely related individuals begin to become problematic for

---

[44] Hoffman, C. L., Workman, M. K., Roberts, N., & Handley, S. (2017). Dogs' responses to visual, auditory, and olfactory cat-related cues. *Applied Animal Behaviour Science, 188*, 50–58.

domesticated dogs? Prior to the Cold War, most dog breeds demonstrated healthy amounts of genetic variation, but over the generations, the desire to have pure bred dogs drove breeders to continue to breed closely related dogs with one another, exaggerating phenotypic and behavioral traits, such as higher drives for seeking, as well as genetic diseases and deformities.

Many of these breeds are specialized for particular work; the herding dogs, for example, have high "seeking" tendencies. According to neuroscientist Dr. Jaak Panksepp, "seeking" is the "basic impulse to search, investigate, and make sense of the environment."[45] But not all of these artificially selected traits have been beneficial for domesticated dogs. A powerful concept in genetics and evolution is "epigenetics," the idea that clusters of genes that control one trait may, in fact, also be involved in the control of other traits, as well. So, as natural selection acts on one trait to favor, or not, the fitness of bearers of a certain trait, other traits may be "carried along" with the trait in question. As we breed for certain traits like appearance, we alter other traits like temperament, learning ability, and even respiration, as Belyaev and Sorokina did with their foxes.

Many dog owners want to have "purebred" dogs, but artificial selection has had harmful effects on numerous dog breeds, due to breeding closely related individuals with one another and selecting individuals for physical appearances without consideration for other associated traits. When closely related individuals produce offspring, this is called inbreeding. With each parent contributing one of its two genes, there's an increased risk the resultant offspring might receive two copies of a recessive, potentially deleterious gene. Offspring borne from closely related individuals may experience physical issues, including compromised immune systems; behavioral problems; exaggerated physical characteristics; and cognitive deficits in comparison to animals that were not inbred.

A 2015 study of 88,635 dogs in California revealed that ten heritable conditions happened more frequently with purebred dogs than with mixed breed ones. These conditions included aortic stenosis (the narrowing of the exit of the left ventricle of the heart), allergic dermatitis, gastric dilatation volvulus (also known as GDV, a condition wherein the stomach becomes stretched and rotated by excessive gas), early onset cataracts, dilated cardiomyopathy (a condition where the heart's ability to pump blood is compromised because the left ventricle is enlarged and weakened), elbow dysplasia, epilepsy, hypothyroidism, intervertebral disk disease (IVDD), and hepatic portosystemic shunt (a bypass of the liver by the body's circulatory system).[46] A 2014 study of 148,741 dogs in England found that purebred dogs had a significantly higher prevalence compared with mixed breed dogs for three different conditions: otitis externa, obesity, and skin mass lesions.[47] Siberian Huskies are predisposed

---

[45] Panksepp, J. (1998). *Affective neuroscience: The foundations of human and animal emotions*. New York: Oxford University Press.
[46] Oberbauer, A. M., Belanger, J. M., Bellumori, T., Bannasch, D. L., & Famula, T. R. (2015). Ten inherited disorders in purebred dogs by functional breed groupings. *Canine Genetics and Epidemiology, 2*(9).
[47] O'Neill, D. G., Church, D. B., McGreevy, P. D., Thomson, P. C., & Brodbelt, D. C. (2014). Prevalence of disorders recorded in dogs attending primary-care veterinary practices in England. *PLoS ONE*. https://doi.org/10.1371/journal.pone.0090501.

to autoimmune disorders, Dachshunds have a high risk of spinal issues, Beagles often have epilepsy, German Shepherds are prone to anxiety and hip dysplasia, Pugs often have breathing and eye issues, Cocker Spaniels are prone to ear infections, and Dobermans are so predisposed to heart conditions that veterinarians often recommend annual screenings. Brachycephalic breeds (those with broad, shortened heads and faces) such as Pugs, Pekingese, Boxers, Chihuahuas, and Bulldogs experience issues such as brachycephalic airway syndrome. Dogs suffering from this syndrome might have small windpipes, or obstructed upper airways – deleterious side effects of being bred for a specific physical appearance. And while the modern bulldog's appearance is widely popular, the majority of them must be delivered by Cesarean section. Prior to the advent of modern veterinary medicine, dogs that would have difficulty giving birth on their own would have been selected against and died out, despite continued efforts to breed them. If all of the moths in the industrialized melanism event had mutations that resulted in lighter wings, instead, the species would have lacked any protective camouflage from predators and would have quickly died out. Through artificial selection from human owners, the Bulldog's head has become so large that it exceeds the capacity of the mother's pelvic canal; natural birth would kill both mother and offspring. But modern medicine enables this breed to remain in the gene pool, and these traits continue to be exaggerated. Morphological traits such as these would have been selected against in the wild – "survival of the fittest" refers not to the most muscular or the most handsome, but to those who are most fit for their environment, such as moths who blend in with bark, finches whose beaks can crack nuts or catch insects, and canines who breathe freely, give birth safely, and have relatively high rates of lifetime reproductive success. Many first-time dog owners don't take into account these genetic considerations when they're selecting a dog; for some, the desire to have a "pure bred" dog outweighs the desire to rescue a dog, even though rescues are also filled with purebred dogs. The dog's intended role in his or her new family often impacts this choice, as well: will they be a show dog? a family pet? a therapy or service dog? a herding dog? a hunting dog? Dogs like Springer Spaniels are wonderful hunting dogs, but they also suffer from a rare genetic condition that results in emotional instability.

In most cases, owners of pure breeds that have known genetic behavior issues (originating from being highly inbred) are aware of them, and their diagnosis, and sometimes, their treatment is based on education from breed groups, breeders, and fellow owners, but occasionally, there are cases where it came as a surprise. Early in my (JCH) career, I naively assumed that *every* owner of a Springer Spaniel knew of "Springer Rage Syndrome," which is a rare syndrome that's most commonly reported in English Springer Spaniels, but can also be seen in other breeds, and manifests in uncharacteristic aggression or sudden attacks. Typically, I would be called in to help modify the aggression that can result from this genetic disorder, which had already been diagnosed by the owner or their breeder. But in one case, a beautiful Springer spaniel named Lucas was exhibiting exactly these symptoms (where the dog's eye's glaze over, and there's a sudden, very violent attack, followed quickly by a calm, "What just happened?" reaction from the dog). It was a surprise to me, and to these owners, that this was a known syndrome, and that they had purchased, for a considerable sum of money, a dog which was now exhibiting this behavior. There was complete disbelief on their part. Sadly, there was no known way to alleviate this dog's symptoms, and his owners ended up returning the dog to their breeder.

Springer Rage, or the more appropriately-labeled "Sudden Onset Aggression" is at least in part a genetic disorder which, with aggressive selective breeding, should be becoming less common, but instead, is now being seen in other breeds like English Cocker Spaniels. I have also seen cases of it in American Staffordshire Terriers, a Miniature Pinscher, and a Vizsla. In my opinion, this syndrome is a sign of poor breeding management, a reluctance to remove affected breeding lines from further breeding, or simply the result of heavy inbreeding. By keeping those individuals who are carriers of Springer Rage in the breeding population, this condition will continue to plague unsuspecting pups – perhaps even worsening over generations as unconscientious breeders continue to breed carriers to carriers.

A more subtle inbreeding effect is creeping into the dog population, at least in the United States, as evidenced by the increased presence of a generalized anxiety disorder. This is likely due to the presence of puppy mills and breeders who are in the business to sell every puppy and make money. There are a *lot* of strange ideas out there about breeding, and the supposed "value" of inbreeding, and a reluctance to rotate breeding studs, et cetera, as it costs more money to purchase and care for more breeding dogs. But the result of a business model that values lower overhead costs over genetic diversity and genetic testing is seen in the rapidly increasing levels of inbreeding in many breed lines. This is in the across-the-board sense, and not necessarily in American Kennel Club-registered and -managed lines; there are lots of responsible breeders taking great care with the management of their dog's genetics, but they cannot meet the demand for dogs in the United States, and there will always be "backyard breeders" and puppy mills who value monetary recompense over being responsible genetic stewards.

This general increase in inbreeding is associated with higher levels of anxiety and anxiety-related aggression. A vast majority (between 80 and 90%) of JCH's in-home cases of aggression were actually the result of fear and anxiety, not dominance, or territoriality, or resource-guarding, and I think that this is associated with inbreeding in many, if not most, of these cases.

WHY INBREEDING IS OUT

Inbreeding issues aren't unique to domesticated animals – or even nonhuman ones. Approximately 10,000 to 12,000 years ago, during the transition from the Pleistocene to the Holocene epochs, three quarters of the animal species on earth went extinct. In North America alone, more than 70% of the continent's megafauna, including mastodons, mammoths, and giant ground sloths, disappeared from the planet forever, with similar extinctions occurring in South America (82%),

Australasia (71%), Europe (59%), and Asia (52%). Subsaharan Africa experienced an extinction of 16% of their mammalian megafauna, including giraffids, the giant hyena (*Pachycrocuta*), and numerous felids, including saber-toothed cats. But a few cheetahs survived this mass extinction to perpetuate the species. This created what's known as a population bottleneck, with all individuals being closely related to one another. Mammals typically share in common 80% of their genes with other members of their species, but each cheetah's DNA is 99% similar to every other cheetah, thus, cheetahs have no genetic heterozygosity. As a result, cheetahs began to exhibit physical indications of their inbreeding similar to the issues with purebred dogs: predispositions to immune disorders and susceptibility to infectious diseases; kinked tails; and an oral dysfunction called focal palatine erosion, wherein the upper palate is eroded and crowds the lower incisors. Because each cheetah shares 99% of its DNA with every other cheetah, they are a particularly vulnerable species. With relatively no genetic diversity and no opportunities to increase the diversity of the gene pool, one disease could easily take out a substantial percentage of the population.

Among humans, inbreeding was once commonplace and seen as a way to preserve certain familial lines – it was the "in" thing to do to literally "keep things in the family." But inbreeding has its downsides, including reduced fertility, congenital birth defects, immunosuppression, certain types of cancer, smaller adult body size, lower birth rate, and higher infant mortality. Families that practiced inbreeding also saw a rise in genetic disorders such as sickle cell anemia, cystic fibrosis, and hemophilia, which are recessive, but occur with higher frequency when both parents are carriers of that recessive gene. Humans, like all animals, receive half of their DNA from their mother and half from their father, but if both parents carry a recessive, deleterious gene, there's a higher likelihood that the resultant offspring will have that recessive gene, as well.

Royal families have practiced inbreeding for centuries; marriages among first and second cousins, nieces and uncles, and other close relatives were encouraged. Because of this repeated inbreeding over generations, rare and recessive genetic disorders became increasingly common. The royal families of Europe have practiced inbreeding to strengthen their lines of succession, ensure the "purity" of their family lines, and secure alliances, but so dedicated was the House of Habsburg to genetic "purity" that a facial deformity was eventually named after them. The Habsburg Jaw, also known as *mandibular prognathism*, was first noted in Holy Roman Emperor Maximilian I, who was born in 1459. After generations of continued inbreeding, the last of the Habsburg rulers, Charles II, was born in 1661. Charles II was referred to as "The Bewitched" because of his physical deformities and emotional and intellectual disabilities. But this wasn't witchcraft: this was the result of an artificially created genetic bottleneck. Charles II's "Habsburg Jaw" was so exaggerated that his tongue lolled out of his mouth and he couldn't chew his food. Not only did he struggle to talk until age four and walk until age eight, but he was also sterile, and died without producing an heir (Fig. 10).

FIG. 10

Charles II.

King Tutankhamun (King Tut, as he became to be widely known) was a teenager who ruled for only a few years, but he is one of the most recognizable historical African figures (Fig. 11). Tut's popular portrayals are far from his tragic reality, though. Tut's parents were likely brother and sister, carrying on a long tradition of "keeping it in the family," and Tut was married to his own half-sister. During his short life, Tut was plagued by debilitating medical conditions, including club foot, the likely result of generations of inbreeding. From King Tut to the House of Hanover, inbreeding was standard practice for royal families. The House of Hanover – and in particular, Queen Victoria's descendants – were stricken with hemophilia, a blood-clotting disease that's caused by a defective X chromosome. Women have two X chromosomes, and they contribute one of these to each child. Women who have the hemophilia gene are usually only asymptomatic carriers of it, while it typically presents in men; thus, for a woman who's a carrier, half of her male offspring will be hemophiliacs and half of her female offspring will be asymptomatic carriers. Hemophilia is painful and potentially lethal, as the blood does not coagulate properly, and sufferers often die young after suffering an injury where their subsequent hemorrhaging could not be stopped. Queen Victoria was a heterozygous carrier of

FIG. 11

King Tut's sarcophagus with Howard Carter, the British archaeologist who discovered Tut's tomb in 1922.

hemophilia, meaning that she had one dominant and one recessive gene for the disease. When she married her first cousin, Albert, who also had one recessive gene for the disease, many of their descendants, including their son Leopold, Duke of Albany, were homozygous for hemophilia. Leopold died at the age of two after he fell and bled internally and two of her grandsons died from hemophilia in their early thirties. Many of Victoria's other descendants, including Prince Alexei Romanov, son of Tsar Nicholas II, had hemophilia.

While far more is known now about the genetics and treatment of heritable disorders, including hemophilia, myriad pernicious genetic issues have arisen among other species throughout history. Despite this knowledge, the practice of inbreeding continues with our companion animals. Among domesticated animals, instances of genetic disorders are most common among dogs, cows, cats, sheep, pigs, and horses – not surprisingly, species that have been highly influenced by human artificial selection.[48] One of the most dramatic cases with horses originated with an American Quarter Horse (AQHA) named Impressive. Born in 1968, Impressive was a champion who lived up to his name, in the show ring and with his fecundity, siring more than 2,250 named foals. Impressive's offspring remained popular to breed to, with many owners breeding back closely related individuals to one another, and soon, he had tens of

---

[48] http://omia.angis.org.au.

thousands of descendants. But then owners of his progeny began to notice odd muscular twitches in their horses that resulted in temporary immobility. The condition was later identified as hyperkalemic periodic paralysis, or HYPP, and after it was traced back to Impressive, it was often referred to as "Impressive Syndrome." Quarter horse breeders are aware of this syndrome, though, and genetic tests now determine whether horses who trace back to Impressive are HYPP-Negative or HYPP-positive, to limit the further spreading of the syndrome in the gene pool. Genetic disorders often arise when a recessive deleterious gene is present in both parents, whether due to intentional inbreeding or mating with closely related individuals when there's a population bottleneck and mating opportunities are limited. Interestingly, among wild animals, instances of genetic disorders are most common among deer, gray wolves, rhesus monkeys, bighorn sheep, and cheetahs, the last of which are particularly genetically vulnerable due to their low incidence of genetic variation.

With the struggle for existence, reproductive rates in the wild are often very high, but most populations remain stable, due to a system of checks and balances with fluctuating resources, predation and hunting by humans, and a variety of other factors, including disease and disadvantageous genetic mutations. A mother wolf might have eight pups, but not all would survive; some would perish due to disease, limited food, congenital abnormalities, predation, or injury; only a fraction would live to bear their own offspring, and only a fraction of those would, in turn, bear their own pups, reflecting lifetime reproductive success for the grandparents. In contrast, a tamed mother dog would likely experience higher reproductive success, especially if her offspring did not exhibit fearfulness of their human housemates.

More than a century and a half after Darwin first noted the immense variation among dog breeds, the American Kennel Club (AKC) currently recognizes 190 specialized dog breeds within *Canis familiaris* – all of them artificially created by humans. There are more than 200 other dog breeds not yet recognized by the AKC, for a total of approximately 400 known breeds. The Westminster Kennel Club categorizes dogs into eight groups: the sporting breeds, including Pointers, Golden Retrievers, Labrador Retrievers, Vizslas, and Irish Setters; Hounds, including Beagles, Bloodhounds, Greyhounds, and Rhodesian Ridgebacks; Working Dogs, including Akitas, Boxers, and Portuguese Water Dogs; Terriers, including Cairn, Bull, and Russell; Toy Dogs, including Italian Greyhounds, Pugs, and Shih Tzus; Non-Sporting Dogs, including Bulldogs, Dalmatians, and Poodles; and Herding Dogs, including Australian Shepherds, German Shepherds, and Old English Sheepdogs, and the "Miscellaneous" Group, which includes breeds that have yet to be officially recognized by the AKC. All of these breeds, so physically dissimilar, yet genetically alike; all created by humans (Fig. 12).

Neither Bulldogs, with their birthing difficulties, nor other brachycephalic dogs, with their breathing difficulties, would be considered "fit" in any natural environment. These animals would likely have experienced limited reproductive success and high predation pressure. And in the wild, aberrant traits such as exaggeratedly large heads that complicated labor or truncated legs that inhibited fleeing from predators would have disappeared from the gene pool just as quickly as they showed up. But

FIG. 12

Different dog groups. *Photographs by Sarah Bous-Leslie.*

in the isolated domain of the human home, potentially deleterious traits can be exaggerated further and further.

With canines' craniums being such striking morphological markers of evolution, Darwin would have been fascinated with some of the trends in canine breeding. Nowhere is inbreeding more visible in dogs than in the brachycephalic breeds. Brachycephaly is neotenization to the extreme – and animal welfare advocates are concerned about the future trajectory of this trend. An organization called "Vets Against Brachycephalism," which is headed by veterinarians, is working to encourage breeders to stop breeding brachycephalic animals. Founder Emma Goodman Milne, BVSc, MRCVS, created the organization after she saw an increase in the incidence of brachycephalic traits and corresponding health issues. Over the past 15 years, she has seen these traits continue to be exaggerated, and she's concerned about the future of artificial selection with these breeds.

Dr. Milne, who became qualified as a veterinarian in 1996, was particularly concerned about breed-related diseases and disorders among her canine patients. "It was immediately apparent in the clinic that brachycephalic animals are the epitome of problems created by man-made and unnatural extremes of body shape," she stated. "In 2004, I was involved in a BBC program about bulldogs, and I've campaigned about brachycephaly ever since."

The trend with brachycephalic traits is continuing toward even more exaggerated characteristics. "Faces are becoming flatter and flatter and the health problems

FIG. 13

A brachycephalic dog and a nonbrachycephalic dog.

are becoming increasingly severe," she explained. "The popularity of these animals, including dogs, cats, and rabbits, is skyrocketing. Persian cats are particularly affected, as are rabbits like the dwarf breeds and lops, which have horrific dental problems. Dogs get almost all of the focus, but we're doing the same thing to many species." (Fig. 13).

Life-threatening levels of disease among brachycephalic animals are higher than their non-brachycephalic peers. "The latest data from the Cambridge BOAS study suggests that over 70% of pugs aged three to seven are clinically affected with respiratory disease," Dr. Milne stated. "What have we done? We're selecting specifically for deformities which make life difficult at best and impossible at worst for these animals. The extremes of body shape are untenable and many breeds are only scraping by through veterinary intervention. Many brachy dogs can't mate or give birth naturally, and this is nature screaming out that it's all wrong."

While the example of the moths with industrialized melanism demonstrates selective advantages, the perpetuation of brachycephalic dogs demonstrates artificial selection gone awry. Evolution is a story of time – of selective pressures, natural or artificial, of mutations, of reproductive success, and of speciation. But the story of the domesticated dog – and of domesticated man – is of rapid, distinct change within relatively few generations, evolutionarily speaking, for canines and for humans, as well. Whether man first came to wolf, or wolf first came to man, neither would ever be the same again. The story of domesticated dogs is just as much a story about us as it is about them – and about countless other people who have spent time with their dogs, just like Alec and Ziva.

Any species' evolutionary history, including its constraints, needs to be taken into consideration, whether you're testing their perceived cognitive capacities or training them to learn new things. Understanding a species' evolutionary history is critical to training, as it considers both their suite of skills and their specific limitations. While all domesticated dogs still belong to the same species, *Canis familiaris*, there are dramatic physical variations among the breeds. It's those dissimilarities that make all the difference for successful training. Just as Darwin's Galapagos finches each had different specializations, so does each dog breed.

You can't expect an insect-eating finch with a grasping bill to excel at eating seeds, which requires a crushing beak, just as much as you can't expect a toy dog like a Pekingese to excel at hunting as well as a retriever would. Similarly, dog trainers have to adjust their training methods and their expectations for each specific breed or type of dog, during instances of unknown lineage. Dog trainers need to understand each dog breed's predispositions, propensities, and limitations to successfully train each individual dog. Some breeds, such as Border Collies and Australian Shepherds, are herding dogs who are "high seek." They have a tendency to keep busy and like to have a job. This makes sense, given their genetic histories; they were bred specifically to herd livestock and outside of this vocation, they thrive in activities such as agility work, which capitalize on this genetic predisposition. Conversely, the Pekingese, a breed that is said to have originated in China's imperial courts, is a "low seek" toy dog that also has a reputation as a breed that's particularly difficult to train. Pekingese are typically content to be low-activity lap-dogs and they likely wouldn't have the propensity to do the agility work that an Australian Shepherd would, nor would they have the same reaction to training methods that worked with herding breeds.

CHAPTER 2

Why tails wag: Umwelts, innenwelts, and canine "guilt"

DOES YOUR DOG REALLY FEEL "GUILTY?"

All apes – including humans – view the world through their own unique lens. From the first ancestor of *Homo sapiens* to today's technologically literate humans, we've all perceived the world with a sensory experience that's uniquely our own. Given our genetic blueprints, we're intrinsically predisposed to do so. Humans excel at communicating with and understanding one another in a variety of sensory modalities, including verbal language, facial expressions and body posture, and pheromones. We're able to detect and respond to minute nuances of human communication, in multiple languages, both verbal and visual. But when we attempt to "see" the world from another species' viewpoint or communicate with them in terms of their sensory perspectives, we falter: we anthropomorphize them. Why is this problematic?

When we anthropomorphize, we attribute human thoughts, intentions, and personalities on nonhumans, most frequently, our companion animals. It's our attempt to understand and empathize with animals, but it falls short of the mark. When people call their dogs "guilty" or participate in "dog shaming," they're anthropomorphizing their canine companions – assuming that dogs would feel as a human might feel if they transgressed. But anthropomorphizing a nonhuman animal actually diminishes them, their evolutionary histories, and their resulting perspectives. If we feel that our dogs have been "bad," we might punish them, but instead of teaching our dogs a "lesson," punishing them for perceived wrongs only teaches our animals that we are mercurial, unpredictable, and unkind. And with our dogs, many still make the error of lupomorphizing them, or viewing the behavior of dogs through wolf-tinted lenses. Both anthropomorphizing and lupomorphizing are flawed assumptions, though – we can't view non-animals in human terms, just as we can't view dogs in wolf ones. Although dogs and wolves share in common 99.6% of their DNA, the two species have key differences. Similarly, chimpanzees and bonobos, two species of large-bodied apes, share in common 99.6% of their DNA; they're also separated by .4% in their genetic makeup. They also share many key differences, including their social structures (chimpanzees are patriarchal, while bonobos are matriarchal), diet (chimpanzees are omnivorous, and hunt, while bonobos are vegetarian), and responses when meeting new social groups (chimpanzees are territorial and will fight violently with rival groups, while bonobos resolve social tension with sexual activity).

Anthropomorphism isn't intrinsically nefarious, though; in the absence of species-specific knowledge, it can help us relate to other animals and has long been used as an engaging literary device. Many critically acclaimed works, including Mark Twain's *A Dog's Tale*, Rudyard Kipling's *The Jungle Book*, Garth Stein's *The Art of Racing in the Rain*, E.B. White's *Charlotte's Web*, and George Orwell's *Animal Farm* (the last of which is particularly interesting, as the animals begin to take on increasingly "human" traits) have anthropomorphized animals. Stories such as these, sometimes rich in irony, satire, or a sense of connection with non-human animals, have resonated with readers for generations, helping people from disparate backgrounds empathize with animals and perhaps have their first introduction to the concept that humans and non-human animals differ from each other in degree, and not kind. While anthropomorphism translates into page-turning prose and is an effective vehicle for empathy, it doesn't yield an accurate picture of other beings' actual thoughts, feelings, experiences, and intentions. It's nice to think that Charlotte the spider would have wanted to help Wilbur the pig, or that Baloo the bear might have wanted to help Mowgli the "man-cub," but this is discounting the animals' distinct perspectives – and what's salient to those species.

That's not to say that animals don't sometimes behave in unexpected ways or form unexpected relationships with one another. Mother dogs have been known to nurse kittens, cats have been known to nurse puppies, and a crow once even adopted an orphaned kitten. But it is the interspecies relationships between a lioness and the animals that would typically be her prey that mystified and fascinated observers more than any other story in recent memory. In 2002, a lioness in the Northern Kenyan Samburu National Reserve adopted half a dozen or more infant antelopes[49] (Fig. 14). The lioness, named Kamunyak (which means "blessed one" in Samburu), was an inspiration to onlookers, who compared her to the biblical story of the lion and the lamb. Kamunyak's forays into "motherhood" came at a great cost: because the antelope calves are wholly dependent herbivores who needed milk and couldn't hide while she hunted, she suffered from starvation while she had each of the calves under her "care." The would-be mother even fought off fellow predators, including other lions, who tried to seize the babies. Some observers anthropomorphized the incidents, believing that Kamunyak loved the calves or adopted them out of loneliness because she didn't have a pride, which lions typically live in. Others assumed that Kamunyak had to have mental health issues to "adopt" another species in such a way. Regardless of the reason, the incidents were certainly aberrant. When one of Kamunyak's adopted calves died from starvation because he couldn't nurse from her, onlookers were shocked that she consumed him. The fairy tale "adoption" was no more. Domesticated cats will "play" with their prey before eating it, but Kamunyak's six forays into "motherhood" appeared to be different. Perhaps we'll never know what Kamunyak was doing, or why. The famed feline and her sister, Dudu, starred in the 2005 film *Heart of a Lioness*, which was filmed from January 2002 through August 2003, but there have

---

[49] https://www.nytimes.com/2002/10/12/world/5-little-oryxes-and-the-big-bad-lioness-of-kenya.html.

FIG. 14

Kamunyak the lioness. *Sketch by Arianne Taylor.*

been no sightings of Kamunyak since 2005. Kamunyak's fate, much like her motives, remains a mystery. Closer to home, we might think that our dogs chew our shoes or our cats pee on our clothing "out of revenge," or that they lick us because they "love us," or that they feel "guilty" if they've eliminated inside the house where they aren't supposed to, but those are our human interpretations of their experiences. They're doing these things for a reason, but not necessarily the one that we're assuming.

There's an intricate interplay between human and canine, a dance whose archaic steps have been performed for tens of thousands of years. And saying that your dog is "guilty," or taking part in "dog shaming," disregards that evolutionary history. So what's really happening when your dog has done something that displeases you and then he has facial expressions and body postures that we interpret as "guilty?" He or she is likely reacting to what you're emoting. Does your dog feel fearful about your reaction to what they've done, based upon what you're emoting and your past reactions about this action? Almost certainly. If you're angry, your dog can tell – they're particularly adept at sensing our emotions. And they're going to react fearfully or nervously in response to your anger when you discover that they've done something that will earn your disapproval. They can sense our anger, even if we think that we've successfully suppressed it, because they have co-evolved with us for millennia. When we see certain behaviors – ears back, lowered posture, wide eyes, tucked tail, teeth showing, licking lips, we see them with our own lens, and we read "guilt," rather than fear or anxiety, because we're attributing human emotions, intentions, and responses onto our dogs. Does your dog feel "bad" that he or she ate the cat food (again) or went through the trash? Maybe – and maybe not. But not in the sense that a human would feel "bad" – we can't impose our morality on our dogs.[50]

So why do our dogs continue to do "bad" things that earn our disapproval? Because they don't see it as "bad" – they see it as "worth the risk." Sneaking the cat food or raiding the trash is a high-risk, high-reward behavior – the food reward is high, but being caught (and punished) is also high-risk. For most dogs, the potential reward often outweighs the potential risk. And dogs aren't alone in weighing costs and benefits for a potential reward. Throughout time, humans have also demonstrated similar risk-taking behaviors. There's ample evidence that high-risk behaviors are as familiar to dogs as they are to man. Archaeological sites reveal that prehistoric man hunted honey and woolly mammoths,[51] which would have been very dangerous and very unreliable. Mammoth hunting wouldn't have always yielded a food reward, whereas trapping small game, foraging for nuts or roots, or fishing would have had more consistent – but far less risky – yields. Honey gathering from wild bees is still practiced today by indigenous people in South America, Asia, Africa, and Australia. While these aboriginal groups rely mainly upon plant material for sustenance, men in the tribe will often seek out high-risk foods. Animal fat and sugar from honey are highly sought after, not only nutritionally, but also because gaining these resources raises their esteem in the eyes of their peers. Honey hunting is actually an ancient human tradition – and one that, along with mammoth hunting, was worthy enough to be documented. There's evidence of both honey hunting[52] and mammoth hunting in ancient cave art. The Cueva de El Castillo (Cave of the Castle), located in Cantabria, Spain, contains the world's oldest known cave art – works that predate both Chauvet Cave and Font-de-Gaume and contain more than 150 different depictions dating back more than 40,000 years. The art depicts images of animals that were important to ancient man, including mammoths, horses, deer, and canines.

Trying to kill large game or ascending 200 feet into a bee-filled tree adds both risk (many are mortally wounded during hunts, die from tree falls, or sustain life-altering injuries) and reward, as the feat of gaining the food is esteemed as much as the food itself. Thus, it's a risk outside of the realm of "morality" – the man ascending the tree and the dog stealing food both feel that the risk is "worth it." And even if the dog or man succeeds only a fraction of the time, that behavior is reinforced just enough to try it again.

A BRIEF HISTORY OF ETHOLOGY

Humans have been studying and documenting nonhuman animals throughout our shared histories. The scientific study of animal behavior, ethology, derives its name from the Greek work *ethos*, which means "character." While it's not a term that's

---

[50] It's possible that dogs have their own version of "morality," as demonstrated by their play behaviors; this will be revisited later.
[51] https://www.naturalsciences.be/en/content/possible-evidence-mammoth-hunting-neanderthal-site-spy-belgium.
[52] http://www.nature.com/nature/journal/v527/n7577/abs/nature15757.html.

in the common lexicon, it has been used for more than two centuries. The term was first used by French zoologist Etienne Geoffroy Saint Hilaire (1772-1844), who was a contemporary of biologist Jean-Baptiste Lamarck (1774-1829), who was the first person to describe a complex (albeit flawed) theory of evolution. But unlike his more famous colleague, who believed that organisms could inherit the characteristics that their parents had acquired during their lifetimes (referred to as "soft inheritance" or the heritability of acquired characteristics), Hilaire believed in the transmutation of species in time, concepts that pre-dated Darwin's theory of evolution by natural selection, which would be published years after the deaths of both Hilaire and Lamarck. While the term ethology was coined in the eighteenth century, animal behavior has been studied – in various forms and with varying levels of "success" – for thousands of years. Man has long wondered why nonhuman animals act in the ways that they do, and researchers from divergent fields have attempted to tackle these questions. Philosopher scientists such as Aristotle (384 BCE – 322 BCE) were early observers of animal behavior; he believed that individuals were born with a *tabula rasa*, or a blank slate: that one did not have innate cognitive abilities, but instead developed throughout the experiences in their lifetimes.

Many early scientists and scientist-wannabes disregarded the experiences of nonhuman animals, perhaps reflecting, in part, Aristotle's *Scala Naturae*. The scale of nature includes all living organisms, with the "most complex" (that would be humans, of course) placed at the top of the complexity pyramid. This paradigm held humans above and apart from the beasts of the earth, influencing the work of later researchers (Fig. 15). Philosopher and mathematician Rene Descartes (1596-1650) discounted the behaviors of animals as reactions and not interactions, stating that animals were automata – unfeeling "robots," essentially – and not sentient beings with their own feelings and perceptions. Cartesian dualism was detrimental to the field of animal behavior for centuries, with vestiges of Cartesian influence still arising in fields as varied as theology and dog training. Through his studies of evolution (discussed in *The Origin of Species*) and emotion (discussed in *The Expression of Emotions in Animals and Man*), Charles Darwin (1809-1882) revolutionized the field of ethology, noting the continuity of behaviors throughout the animal kingdom. Under Darwin's encouragement, his protégé, Canadian evolutionary biologist George Romanes (1848-1894) studied animal intelligence with a method called "anecdotal cognitivism." This method was viewed as anthropomorphic (and thus unscientific), and never gained traction. British evolutionary biologist Julian Huxley (1887-1975) focused on behaviors that occurred in all members of a species. Working with domesticated chicks in 1873, British researcher Douglas Spalding (1841-1877) was the first ethological observer to note the occurrence of a phenomenon later referred to as "imprinting," in which a young animal will become attached to the first animal they have interactions with. Austrian ornithologist, ethologist, and zoologist Konrad Lorenz (1903-1989) was considered to be one of the founders of modern day ethology, and shared the 1973 Nobel Prize in Physiology or Medicine with Karl von Frisch and Dutchman Nikolaas (Niko) Tinbergen (1907-1988), who was also considered to be a Founding Father of ethology. Working primarily with avian species, including geese, Lorenz

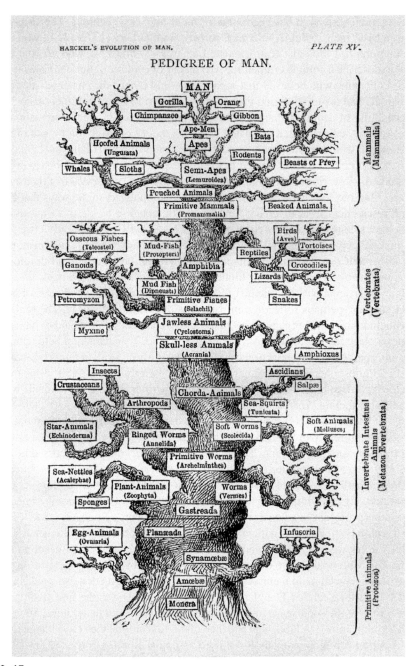

FIG. 15

The scale of man.

investigated the concept of imprinting, finding that newly hatched birds would form attachments to the first moving objects that they viewed. Lorenz' 1949 book, *Man Meets Dog*, includes his own personal anecdotes with dogs, as well as insights into their behavior.

Tinbergen is perhaps most well known for his "four questions," a concept that was built upon Aristotle's "four causes" and provide explanations for behavior. The four questions are causation, which pertains to the stimuli that produced the individual's response, survival value, and how the behavior contributes to the individual's survival and lifetime reproductive success, ontogeny, and how the behavior develops during an individual's lifetime, and evolution, and how the behavior became a part of the species' repertoire. Causation and ontogeny are proximate mechanisms, while survival value and evolution are ultimate mechanisms. The former describes responses to immediate factors, while the latter describes the purpose of the behavior.

While Darwin, Lorenz, and Tinbergen provided a rich and robust foundation for the field of ethology, the field drew from a wide variety of disciplines, including psychology, biology, and ecology. With so many divergent backgrounds in this new field, there was dissent among the scientists' methods and results. American psychologist John B. Watson (1878-1958), author of the 1924 book *Behaviorism*, believed that psychology was unrelated to mental activity, but could instead be explained with physical actions. Watson believed that human children could – and should – be raised objectively and devoid of affection. One of Watson's students, Burrhus Frederic (who unsurprisingly preferred to go by his initials, B.F.) Skinner (1904-1990), believed that behaviors were dependent upon the results of prior actions, with adverse consequences yielding fewer repetitions of a behavior than favorable ones. When consequences were favorable, the likelihood of repetition increased, and Skinner coined this as the "principle of reinforcement." Skinner used "operant conditioning" to manipulate an animal's behaviors, using either punishment (the adverse consequence) or reward (the favorable consequence) to influence the rate of a particular behavior. Russian physiologist Ivan Pavlov (1849-1936) similarly focused on classical conditioning, noting that dogs would have a physiological response in anticipation of a food reward.

Kenyan paleoanthropologist Louis Leakey (1903-1972) took a different approach: he worked with three women who had expressed a keen interest in animal behavior (Jane Goodall, Dian Fossey, and Birute Galdikas, collectively referred to as "the trimates") and a willingness to go into the field in Africa and Asia to observe the behavior of primates in free-living habitats. Studying chimpanzees, gorillas, and orangutans, respectively, the collective observational findings of Goodall, Fossey, and Galdikas would revolutionize the field of ethology almost as much as Darwin's work had steered the course of evolutionary research. Modern ethology continues to build upon the work of these diverse researchers, utilizing a multi-disciplinary approach, including observations, physiological responses, and the understanding that different species will have divergent sensory experiences.

To understand behavior, you have to integrate multiple levels of influence. Behavior is the ultimate result of a series of hierarchical influences, from more-remote,

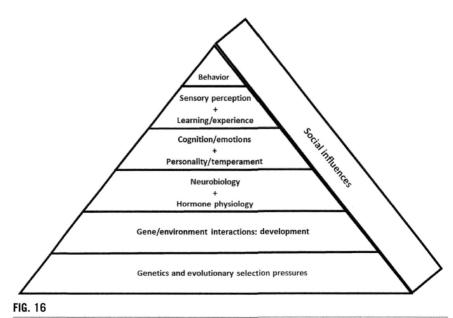

FIG. 16

A graphical depiction of the levels of integration involved in behavior.

evolutionary, population-level influences on genetics, through developmental interactions between genes and the environment, to the development of the brain and hormonal control systems. More immediately, behavior is the result of established cognitive capacity, including emotional responses and personality dimensions, which process sensory perception as well as immediate environmental influences through learning and experience. Social interactions influence behavior at every level in this hierarchy, from evolutionary selection pressures to development of the individual to personality, and of course, in an immediate sense as part of the animal's environmental interactions (Fig. 16).

INSIDE OUT

To understand what's going on with our companion animals, you need to understand the concepts of "umwelt" and "innenwelt." German biologist Jakob von Uexküll[53] created these concepts in the early 1900s to illustrate how different organisms within the same ecosystem not only interpret signals differently, but find some signals salient and others inconsequential. One's umwelt is the part of the world that an organism can detect – it's how the world appears via their unique perceptual systems. One's innenwelt is their sense of self – of "I." Within one meadow, there could be many different species of animals, all experiencing the environment from their own umwelts.

---

[53] http://www.zbi.ee/uexkull/link.htm.

For the bee, the world is viewed through an ultra-violet lens; for the tick, who is both deaf and blind, the odor of butyric acid is all-important, and the world is experienced through temperature; for the moth, light and odor are both attractants; for the bat, the world is experienced through echolocation; for the rattlesnake, the world is experienced through infrared or "heat" vision; and for the canine, olfactory cues are particularly important. Ziva smelled the strange man long before Alec knew that he was there, and alerted Alec. And for the human, there's a high reliance on visual cues and the propensity to create patterns, especially faces. This psychological phenomenon is called pareidolia and involves a stimulus that the mind perceives as having a familiar pattern. Humans often see "faces" in objects that don't contain them, such as tree branches, food, or clouds. But faces are salient to us – and we're "primed" to see them. Whether it's Jesus in toast, Donald Trump in butter, or Madonna in a pancake, we "see" what's salient to us. So too does a bee, a snake, or a bat perceive what's salient to them. And each organism, supposedly blissfully unaware of any other reality, likely assumes that their umwelt is "reality." And why wouldn't they? If a flower looks yellow to us, or has a light fragrance, why would we think it would be "interpreted" in any other way? But for the bee or the dog, that flower would be experienced in a range of color and scent that we could've never imagined before modern technology gave us a window into their worldviews (Fig. 17).

FIG. 17

The umwelt of ticks, moths, and bees vary significantly from each other and from *mammals. Sketch by Arianne Taylor.*

Human infants experience similar difficulties when trying to understand different perspectives. A young human child doesn't understand object permanence or theory of mind (ToM); they think that someone else can only see what they can see. And this is to be expected – they haven't yet learned that there are perspectives and experiences outside of their own. This phenomenon was demonstrated in the movie *The Truman Show*, where the main character, Truman, is actually part of a lifelong reality TV experiment. He doesn't question a life that would be considered "unconventional" by the standards of outsiders, because they're the only standards that he has ever known.

While umwelt refers to one's environment, innenwelt refers to an organism's "inner world." French psychoanalyst Jacques Lacan used these terms to demonstrate the interplay between self and environment. It's a concept that can be difficult to grasp, even on a same-species level, so it's no wonder that it's so hard to try to "walk a mile in the shoes" (or paws, in this case) of another species. Think of how disorienting it is to look at a world map that's oriented differently from what you're accustomed to – or if the land portion and the sea portions are reversed. Or imagine if you're trying to discern Chinese characters, but you've only ever learned a Germanic language. Trying to understand another species' umwelt is even more unfamiliar.

Humans have their own evolutionary baggage and we view the world through human perspectives. With the exception of the spider monkey, who has four-fingered hands, all primates have hands with five fingers – they have retained a primitive mammalian trait called pentadactylism. Apes, including humans, chimpanzees, bonobos, gorillas, orangutans, siamangs, and gibbons, have forward-facing eyes, lack an external tail, are highly encephalized, and have the same dentition pattern. Apes rely more upon their vision than they do their sense of smell. Canines, in comparison, rely more upon their sense of smell than they do their vision – according to some estimates, a dog's sense of smell is as much as tens or even hundreds of thousands of times better[54] than a human's.

THE MATRIX IS REAL

For bees, the world is painted in ultraviolet, and for dogs, the world is just as different – but it's built with a cityscape of smells that begins several feet below the horizon of a human's perspective. Canines have a three-dimensional sense of smell – they can discern so much from scent that they have an olfactory worldview that's as foreign to our sense of smell as a bee's ultraviolet worldview is to our sense of sight. Think *Alice in Wonderland* meets *The Matrix* – that's just how foreign dogs' sensory world would be to us, if we tried to truly walk a mile in their paws.

What can dogs tell from scent? They can detect who was there, and whether they were male or female; they can smell drugs and explosives, even if they're buried or

---

[54] www.pbs.org/wgbh/nova/nature/dogs-sense-of-smell.html.

obscured by other scents; and they can smell cancer cells on their human companions. By some estimates, dogs can follow a scent trail for hours or even days.[55]

Dogs' noses don't just differ from ours in function; they differ in form, as well. Humans have six million olfactory receptors in their noses, which might sound like a lot – until we compare that with the 300 million olfactory receptors our dogs have. While apes are highly encephalized, dogs' brains are highly specialized – the portion of the canine brain that's devoted to olfaction is 40 times greater than the equivalent area in the human brain. When dogs inhale, the air is separated into two distinct routes: One for respiration, which passes to the pharynx and the lungs, and one that goes into the dog's olfactory region. Dogs can move their nostrils independently, as well; humans, on the other hand, breathe and smell through the same nasal passages. Dogs don't just smell with their noses, however – they also have a secondary olfactory organ called the vomeronasal organ (colloquially known as the Jacobson's organ[56]), which can detect pheromones and tell them important information about another being, including their reproductive status.

BUT WAIT, THERE'S MORE

While dogs have coevolved with humans for millennia, they last shared a common ancestor as long as 100 million years ago. As a result, dogs have a different blueprint than we do. Their sensory experience and their viewpoint (including standing around one to two feet, on average, vs. five to six feet for a human) is uniquely their own. And in addition to a superior olfactory sense in comparison to humans, dogs also have hearing that's superior to ours.[57] Dogs can hear sounds at higher pitches and can detect a wider frequency range (67-45,000 Hz, compared to humans, who have a range of 64-23,000 Hz). Humans and dogs also interpret visual data differently. Humans typically have three types of cone cells in their eyes, while dogs only have two. As a result, dogs have difficulty distinguishing between green and red, seeing both as variations of yellow. This is similar to how approximately 1 in 8 men and 1 in 200 women are "color blind."[58] Dogs do have superior vision to humans at dusk and at dawn, however, and also recognize moving objects better than humans do, just as Ziva could see the moving wolves before Alec, but humans have superior vision up-close and at a distance. It could be said that dogs have low-resolution vision in comparison to humans. Thus, a dog's world is constructed more with smells and sounds than with the other senses, while humans rely heavily upon their sense of sight. These differences and similarities all shape how we view and interact with the world.

So much goes "undetected" from one organism to another, simply because we have different umwelts. But just because dogs can smell far more than humans and

---

[55] http://www.vsrda.org/about-vsrda/using-air-scent-dogs.
[56] https://www.whole-dog-journal.com/issues/7_11/features/Canine-Sense-of-Smell_15668-1.html.
[57] https://www.lsu.edu/deafness/HearingRange.html.
[58] https://nei.nih.gov/health/color_blindness/facts_about.

have a greater range of hearing, that doesn't mean that humans experience an "absence" of smell or sound; and just because bats use echolocation and bees see ultraviolet, doesn't mean that we have an absence of these sensory experiences, either. The portion of the electromagnetic spectrum that's visible to humans is less than a ten trillionth of the entire spectrum – but we don't miss the rest of the spectrum, just as a person who has always been blind doesn't miss having vision, because that's all they've ever known. According to animal behaviorist Dr. Temple Grandin, animals not only perceive the world differently, but conceive it differently, as well. They think – and remember – in pictures and with associations.

"An animal's sensory world is different from ours," Grandin explained. "Dogs hear high-pitched sounds that we can't. Birds can see ultra violet. Think about a dog's sense of smell. An especially talented wine steward could identify 3,000 wines by smell. Most people can't begin to do that, but that's a good way to begin to understand a dog's sense of smell."

According to Grandin, when a dog sniffs a tree, he's receiving a wealth of important information. "It's like his smellivision," she explained, "because he can tell who was there before him. You have to try to understand the sensory details he's experiencing."

FACS AND FICTION

Think of when you're driving on a busy freeway – especially if you've recently had an accident. You're going to be tuned in to the vehicles around you, including their speed and placement in the lane. If one vehicle moves, even almost imperceptibly, toward your lane before signaling, you'll notice. The driver then indicates that they're signaling, but they've already hinted that they're going to be moving in that direction – in the microsecond before they signaled, the car drifted in the direction that they intended to travel. As a fellow driver, you tune in to this because it's particularly salient to you – to stay safe, you need to maintain a certain distance from other vehicles, and a car coming into your lane is dangerous.

Our faces make similar movements that betray our intended actions and feelings. Facial expressions can be just as important to other people and to our animals, as well, who can tune in to our emotions without our awareness. These fleeting, almost imperceptible facial microexpressions last only 1/25 to 1/15 of a second, but within that fraction of time, they reveal a person's actual emotional state, before he or she can conceal it. So who's good at reading microexpressions? Con men and companion animals – and dogs, in particular.

Researchers from the University of Portsmouth recently found evidence that our dogs preferentially use facial expressions when we are oriented toward them. The researchers studied 24 dogs, aged one to 12, of various breeds. Using an anatomically-based coding system called "DogFACS," which measured facial changes and underlying muscle movements, the researchers found that if the dogs were looking at something other than a person's face, even at a value-laded item like food, they exhibited fewer emotionally-related facial expressions. One of the most commonly exhibited expressions was eyebrow raising, which gives the appearance of larger, so-called "puppy dog" eyes. Lead researcher Dr. Juliane Kaminski believes that these behaviors developed over time during the domestication process. She stated: "We can now be confident that the production of facial expressions made by dogs are dependent on the attention state of their audience and are not just a result of dogs being excited...The findings appear to support evidence dogs are sensitive to humans' attention and that expressions are potentially active attempts to communicate, not simple emotional displays."[59]

So why is a facial expression like "puppy dog" eyes so important? Kaminski hypothesized that early on during the domestication of dogs, humans preferentially selected early dogs who exhibited neotenous facial expressions over dogs that did not. Neoteny, also called pedomorphism, is the retention of juvenile features in an adult animal. Over multiple generations, the foxes that were selectively bred by Belyaev and Sorokina exhibited both "puppy-like" morphological traits and friendlier dispositions. Neotenous physical appearances seem to co-occur with more human-friendly temperaments with canines, so it's difficult to ascertain whether early dogs were being selected more upon their appearance or on their behavior, but many modern domesticated dog breeds show particularly neotenous traits, including the Cavalier King Charles Spaniel, the Japanese Chin, and the Pomeranian. Kaminski believes that puppy-like behaviors and appearances evolved as a byproduct of selecting against aggression, with highly social, expressive dogs having a selective advantage over less social ones.

There's something to the phrase "you begin to look like your dog" – often, our canines are selected for their neotenous, "almost human" characteristics. Humans' obsession with neoteny goes beyond wanting to have companions with babyish traits – people want to capture that "younger" look and lifestyle, too. In his book *Juvenescence: A Cultural History of Our Age*,[60] Stanford professor Robert Harrison uses neoteny as the point of departure for a youth-obsessed culture, noting that modern advances in health and medicine have brought about a biocultural paradigm shift. Chasing after youth is exemplified by the multitude of anti-aging, anti-wrinkle creams available on the market – products that purport to help you "defy the aging process" and make you "look younger in weeks." It's exemplified in our use of hair dyes to hide gray hair, plastic surgery that hides wrinkles, opens up eyes, plumps lips, and lifts entire faces, vitamins to keep us healthy and "combat aging" (as if that's possible), and in the endless supply of cosmetics to make our eyes look larger and our lips look redder and fuller. It's clear in our adulation of babies – human and otherwise. And it's perhaps clearest in our choice of companions. Kaminski believes that early dogs exploited humans' appreciation of neotenous characteristics, lending support to the complexity of the domestication process of dogs. We love puppy dog eyes – and our dogs love to make these expressions.

While the Dog Facial Action Coding System (or DogFACS) does not detect emotions, it does identify and code facial movements in dogs, allowing researchers

---

[59] Kaminski, J., Hynds, J., Morris, P., & Waller, B. M. (2017). Human attention affects facial expressions in domestic dogs. *Scientific Reports, 7*(12914).
[60] Harrison, R. P. (2014). *Juvenescence: A cultural history of our age.* University of Chicago Press.

to code facial expressions. According to Darwin, emotional facial expressions are universal and are the product of evolution. This hypothesis has been bolstered by studies with blind people who have no visual reference for others' facial expressions. So, too do nonhuman primates have universal facial expressions. Our closest cousins, the chimpanzees, are often seen "smiling" on greeting cards, but it isn't a "smile" at all; it's a fear grimace – and it's universally expressed. This innate response is demonstrated in humans who are reacting to something painful or stressful – like the moment before a car crash or putting one's hand on a hot burner. Nonhuman apes, too, make this expression when they're fearful, and training techniques to elicit this response for commercial use are highly controversial – and unethical, as well.

Microexpressions were first discovered in 1966 by Haggard and Isaacs, who found "micromomentary expressions." Later, Paul Ekman[61] built upon Darwin's hypothesis, finding that many emotions were universally expressed among humans, including sadness, surprise, fear, happiness, anger, and disgust. William S. Condon spearheaded research into social interactions measured at fractional intervals of each second. During the 1960s, he gained fame for analyzing a four and a half minute film at intervals of 1/25 of a second, with each frame accounting for one interval. Microexpressions include masked expressions, where actual expressions are concealed by false expressions; neutralized expressions, where actual expressions are suppressed; and simulated expressions, which are expressions that aren't accompanied by real emotions.

DogFACS wasn't the first tool to measure and catalogue facial expressions, though. In the 1970s, Ekman helped develop the original Facial Action Coding System, or FACS, to catalogue human facial expressions. FACS can describe every observable facial movement that occurs with every emotion. Dogs use microemotions too, but having co-evolved with humans for millennia, they don't need a FACS to tell them fact from fiction. Dogs excel at reading us – from our scents (are we healthy, or stressed, or do we have abnormal cell growth?), to our emotions (are we mad at them, or are we happy or excited?) to our facial expressions – even the ones we try to hide. Many people use their dogs as a gauge of other peoples' character – and perhaps with good reason.

HOW TO READ A DOG

How often have we wanted to know exactly what our pets were saying, in both their verbal and nonverbal communication? The popular television show *Lassie* was entirely based upon one canine heroine's series of whines and whimpers, to which her people always responded, whether they understood what she was saying or not. Disney's *The Cat From Outer Space* featured an other-worldly Abyssinian who could telepathically speak English, the television program *Mr. Ed* featured a talkative palomino, and more recently, Pixar's *Up* included a dog named Dug who had a dog-to-human language translator affixed to his collar, affirming that dogs do, in fact, idolize

---

[61] https://www.paulekman.com/micro-expressions.

humans. Throughout the film, Dug, via the translation device, communicates with humans, emitting astute announcements such as, "I was hiding under your porch, because I love you. Can I stay?" and "Oh boy! Oh boy! A ball! I will go and get it, and then bring it back!" But he is best known for his single-word utterance, "Squirrel!" Dug's translation device wasn't wholly science fiction, though; recent research shows that while we don't necessarily know the meaning of dogs' "words," certain sounds that dogs make do contain certain meanings for them, and the amplitude and pitch of each bark can carry a lot of information with it, including the size of the barker.

Playback studies of vocal communication have been performed with multiple primate species to determine if they have referential communication, for example, if they use the same vocalization for "eagle" that they do for "leopard." This could be problematic, if they did, as the proper response to an avian predator would be to flee from the treetops, while the proper response to a terrestrial predator would be to flee to the treetops. The results of these studies have been mixed, with some results revealing that there were different sounds (words?) for aerial predators versus terrestrial ones.[62] Some studies also revealed referential labeling that provided not only the predator "category" (aerial or terrestrial), but information about the predator's distance, as well.[63] Playback studies with our closest living relatives, the chimpanzees, have revealed referential vocal communication with "food grunts," with different grunt types corresponding to different types of food.[64]

Similar playback studies have been performed with domesticated dogs, revealing that dogs do use context-specific referential communication.[65] The dogs in the latter study used context-dependent vocalizations during agonistic food-guarding encounters with their peers that differed from the growls used in the play context and with two different types of agonistic growls. Over the course of multiple playbacks, the food-guarding growls deterred dogs from absconding with an apparently unguarded bone more effectively than the play growls or the other agonistic growls. It appeared as though this specific growl meant, "This bone is mine," or some variation thereof. Now if only we could extend the results of these experiments to the entirety of canine communication!

Since we can't understand a species from their own umwelt, and lacking dog-to-human translation devices, how can we begin to understand where dogs are coming from and what they're trying to communicate? We can learn our dogs' unique individual communicative gestures and expressions, but unlike our dogs, who expertly read their humans, we need a "how-to" guide to understand the subtle signs they give to us and to one another. The ethogram, which catalogues the entire behavioral repertoire of a species, is a pivotal tool, both for ethological research and for companion animal

---

[62] Fichtel, C., & Kappeler, P. (2002). Anti-predator behavior of group-living Malagasy primates: Mixed evidence for a referential alarm call system. *Behavioral Ecology & Sociobiology, 51*(3), 262–275.
[63] Zuberbuhler, K. (2000). Referential labeling in Diana monkeys. *Animal Behavior, 59*(5), 917–997.
[64] Slocombe, K., & Zuberbuhler, K. (2005). Functionally referential communication in a chimpanzee. *Science Direct, 15*(19), 1779–1784.
[65] Farago, T., Pongracz, P., Range, F., Viranyi, Z., & Miklosi, A. (2010). This bone is mine: Affective and referential aspects of dog growls. *Animal Behavior, 79*(4), 917–925.

welfare. An ethogram catalogues an animal's behavior, comprising the placement, configuration, and movement of body parts and behavioral contexts, including locomotion and travel. The behaviors contained within an ethogram are typically defined as objective and mutually exclusive, thus making them individually recognizable to any observer. Ethograms can include a wide range of information, such as the social context, gestures, facial expressions, vocalizations, initiators, and recipients (Fig. 18).

An example of an adult, nonreproductive dog ethogram (without sexual and infant care behaviors)

Low activity
- Lying — torso on the ground
- Sitting — rear legs on the ground, front legs extended
- Standing — torso off the ground

Moderate activity
- Exploring — low-level locomotion, sniffing and visual investigation
- Social — low-level, casual investigation of other dogs and play

High-intensity activity
- Trot — lower speed locomotion
- Gallop — intermediate speed travel
- Run — full-speed travel

High confidence
- Tail up — self-explanatory
- Ear up — ear fully erect and displayed

Moderate confidence
- Neutral ears — ears to the side
- Neutral tail — tail at or below horizontal

Low confidence
- Ears back — ears flat to the scruff/neck
- Tailed tucked — self-explanatory

Low-intensity social
- Greeting — approach to another dog with intent to interact
- Nose touch — self-explanatory
- Inguinal sniff — self-explanatory

Moderate-intensity social
- Rally — multiple dog participation/recruitment

High intensity social
- Play bow — front legs stretched forward, front body lowered
- Stalking — stiff-legged gait, tracking and following another dog
- Play biting — biting without intent to break the skin, grasping only

Flight
- Escape — self-explanatory
- Leave — casual, low-intensity departure from social situation

Mixed fight and flight
- Whirling — spinning movement to escape attack
- Defense snap — an "air snap" used as a signal of arousal
- Snarling — characteristic vocalization in mixed-state behavior

Fight
- Lunge — self-explanatory, with intent to bite with damage
- Wrestle fight — self-explanatory, with intent to bite

Adapted from the work by Jane Packard and Sonia Alvarez, based on the work by Roger Abrantes (1997) (Dog Language) and Erik Zimen (1975) (The Wolf)

FIG. 18

A canine ethogram.

An example of a simple dog ethogram
For use in a shelter-based research project (i.e. for singly kenneled dogs) by J.C. Ha

Movement
    Pacing
    Jumping
    Hopping
    Walking

Postures
    Laying
    Sitting
    Stand: all fours
    Stand: rear legs

Location in kennel
    Front
    Middle
    Back

FIG. 18, CONT'D

For example, a dog's ethogram would include the social context, such as feeding, greeting, grooming, or play; the placement of each body part, including the head, tail, torso, and legs; accompanying vocalizations; the configuration of each body part; and the movement of these body parts. Tail positions would include more than 90 degrees, equal to 90 degrees, and less than 90 degrees, with each placement indicating possible differences in arousal level and context. Tail configuration would include straight or curled. Tail movement would include a tail wag – repeated side to side movement – or a motionless tail. Each observed body position would have a specific name, for example, "ventral lie down" would entail the abdomen making contact with the substrate; "dorsal lie down" would entail the back making contact with the substrate; and "lateral lie down" would entail the side making contact with the substrate. The descriptions of these behaviors would be specific enough to allow independent observers to reliably agree with one another on what they were seeing.

From nose to posterior, a detailed ethogram would enable observers to reliably document and decipher the behaviors of a species. For example, what would it mean when a dog was wagging their tail? The communicative intent could be deciphered using the angle of the tail, the type of movement, and other behaviors that co-occurred with the placement and movement of the tail, including body posture, facial expressions, and vocalizations, or lack thereof. For example, a rapidly wagging tail could indicate excitement, a broad tail wag could indicate a dog who is being friendly, a slight tail wag could indicate a greeting or interest in a highly desired item (especially when paired with begging behaviors, such as "puppy dog eyes," sustained eye contact, and gaze following) and a broad, sweeping tail wag that incorporates movement of the hips could indicate greeting an individual that is special for that dog. If the tail was erect, between horizontal and vertical, and stationary, this could be a sign of alertness or aggression. A tail that was erect and curled over the dog's back could indicate confidence, while a lowered, limply moving tail could indicate depression, pain, or illness. If the tail was horizontal, relaxed, and pointing away from the dog, this could indicate interest. A tail that was at approximately 45 degrees and wagging slowly could indicate uncertainty or insecurity, while a tail tucked between the hind legs could indicate fear, timidity, or insecurity. Sometimes, tail wagging appears to be involuntary, as in when a dog sees a highly valued food item and wags their tail with excitement. Interestingly, chimpanzees have a food grunt vocalization that's restricted to the eating context, and chimpanzees often can't help but make this audible announcement of gustatory anticipation, even if they're trying to sneak food.

When Romeo the ill-fated wolf first met his "Juliet," the moment was captured on film. Dakotah, as she was known to her friends, was standing nose to nose with the larger canine. Her head was up, her ears oriented forward, and both canines' tails were placed straight out. The dogs appeared to be curious about each other, but not intrusive; they were both assertive, but not aggressive. Let's take a closer look at the behaviors that Ziva was exhibiting as she walked with Alec to the blind, beginning with her tail. Ziva's tail placement, configuration, and movement indicated her emotional state and her attempts to communicate with Alec and with the strange new dog. When she was first waiting with Alec, her tail was wagging gently against the ground, indicating that she was relaxed and happy. Later, when she smelled wolves nearby, she was on high alert and her tail was at a 90-degree angle. When Ziva met the new dog, their tails communicated very different things: he wagged his tail slowly, but he also cowered on the ground; thus, his tail wagging could indicate uncertainty or an attempt at self-calming behaviors. Conversely, Ziva's tail was also wagging slightly, but her rump was raised in the air, closely resembling a play bow, or invitation to play. Ziva was being friendly and greeting the other dog. When the other dog left, he trotted away with his tail tucked, indicating that he possibly felt insecure, perhaps as a result of the reprimand that he had received from his owner. The placement, configuration, and movement of the tail might initially go unnoticed by a casual observer; those who have spent any amount of time with a dog could likely discern the implied meaning that the tail was conveying on a subconscious level. We don't often catalogue each gesture in a social interaction – it's time-consuming and tedious. But we code behaviors on a subconscious level every single day in our social interactions with people and with our pets, as well, with varying levels of success. "First impressions" of an individual are often accurate because we're cueing into subtle, subconscious communicative signals. Not everyone will cue into subtle social interactions, though; and not everyone is equally adept at observing and interpreting behaviors. Sometimes, fear might be misinterpreted as aggression, and a fearful dog might be reprimanded for simply demonstrating that they're insecure about a situation. Similarly, a dog that's exhibiting threat or alert behaviors might be misinterpreted as a dog that's friendly and interested in another dog, resulting in an altercation between two new dogs that observers swear was "out of the blue" when there were actually indications from one or both dogs. When we do know how to document and decipher dog behaviors, gesture by gesture, though, they paint a very clear picture of canine communication.

A dog's eyes and ears are also strong indicators of their communicative intent. Is the dog staring straight on, averting their gaze, blinking frequently, gazing about, or changing their gaze between a high-value item (such as food) and a human recipient? Straight-on staring can indicate a threat, averting their gaze can indicate discomfort or uncertainty, frequent blinking can be a self-calming behavior, and alternating one's gaze between a desired item and a person is classic "begging" behavior. After Alec gave Ziva a piece of jerky, she looked from the bag containing the rest of the jerky back to Alec's face, affixing her eyes on his, and was "rewarded" for her begging with another piece of jerky. When Ziva submissively exposed her stomach to the new dog, the whites of her eyes were showing, indicating that she was timid and uncertain about this stranger. Whether eyes are half shut or wide open, staring or glancing, blinking a lot or squeezed tightly shut, these are all behaviors that can be used to discern a dog's emotional state. And when a dog averts their gaze with their head oriented toward the floor after a perceived transgression, they aren't feeling guilty; they're likely nervous or fearful and anticipating the response that will come from their human.

A dog's ears can provide subtle cues to their emotional states, as well. If the ears are flattened against the head, they can indicate either submission (in the absence of bared teeth) or anxiety (in the presence of bared teeth). Ears that are forward-facing and erect indicate a dog that's alert, while ears that are splayed out and pulled back can indicate fear, indecision, or submission. Ears that flicker back and forth can indicate indecision, while ears that are flopped over or "inside out" can indicate a dog that's very relaxed. When Alec told Ziva that she couldn't have a third piece of jerky, her ears flicked back and forth rapidly, indicating her indecision over choosing to accept "no" or trying to beg for food once more, as it worked the first time, but didn't the second. Later, as Ziva searched the air with her muzzle, her ears moved forward, as well, indicating that she was on the alert. When she first met the new dog, her ears were back, but her teeth weren't bared, indicating her submission to the larger new dog.

In addition to vocalizations, yawning, and licking, a dog's mouth can provide a lot of information about their emotional state. A dog with a relaxed, partially open mouth is relaxed, while a dog with their lips curled back and their teeth exposed is fearful or aggressive. When Ziva was walking alongside Alec after receiving her treats, she had a relaxed jaw. Showing one's teeth as a "warning" is a behavior that's exhibited across multiple animal species, including primates, felines, and canines. When the new dog met Ziva, he curled back his lips and bared his teeth at her, to which Ziva immediately assumed a submissive position. Dogs often use vocal communication, but non-vocal mouth communication, including licking and yawning, is important, as well. Yawns can indicate an attempt to calm oneself – both humans and canines will yawn during moments of stress or uncertainty. Yawns can also be a reaction to a human or another dog yawning, an attention-seeking behavior, or an attempt to wake themselves up. Alec yawned, apparently because he was tired, and Ziva reciprocated a yawn, likely because Alec had yawned first. Depending upon the behavioral context, licking can be affinitive, as in greeting a familiar dog or person, self-directed, as in grooming, attention seeking, or an attempt at self-calming. Licking or showing one's tongue can be an affinitive or calming gesture; dogs might lick the air or ritually lick random objects when nervous. Among horses, licking is also an indication that the animal is being submissive: many horsemen interpret this gesture as, "I'm an herbivore; I mean you no harm."

The position of the head can also indicate a dog's emotional state. A head that is downward-facing is indicative of a dog that's likely nervous, shy, or submissive, while a raised head can indicate a dog that feels threatened or aggressive. A dog who approaches another dog with their head tipped to the side is indicating that they are no threat. When Ziva sensed the wolves, her muzzle was raised; she was on the alert. When the new dog met her, his head was up, and combined with bared teeth, he appeared to feel threatened or aggressive. When the dog departed, his head was low and downward facing, indicating his likely submission or uncertainty with his owner.

The placement of the body is also indicative of a dog's emotional state. A dog can be standing with their body lowered, indicating submission, or upright and stiff-legged, indicating aggression or a perceived threat. Their body can be in a "play bow" position, with their rump raised into the air and their front legs along the substrate, or they can be in a "submissive" position, with their dorsal on the substrate and their stomachs exposed, as Ziva did to the new dog (Fig. 19). Sitting can indicate relaxation or comfort in the presence of other dogs and people, rolling over on the side while avoiding eye contact can indicate submission, and rolling onto the substrate and then rubbing their back, hips, and shoulders against it indicates a dog who is very happy. The body can also show other indications of emotion, including piloerect hair (bristling of the hair, typically on the neck and along the back).

All of these behaviors are subtle and, given the co-evolutionary history of humans and canines, natural to see and to read about. But when they're painstakingly divided into their individual components, they can help observers understand "why" they may have had the impression that a dog was friendly or unfriendly, alert or relaxed, excited or afraid.

Using an ethogram can be particularly helpful for troubleshooting problem behaviors. There are many in-home behavior cases which, at least in part, stem from

FIG. 19

A play bow. *Photograph by Sarah Bous-Leslie.*

the owner's lack of understanding of canine body language. JCH has had a few such cases. The classic examples are cases where I'm called to a home to assist in dealing with a dog-aggressive dog, at a not-insignificant cost to the owner. After taking a full history and observing the dog in social settings, I simply have to break it to the owners that they have…a very playful dog! She's perhaps a "Rough and Tumble Player," as I call them, but there's no sign of aggression… check please, thank you very much!

The concept of an ethogram is important when talking to owners about body language. JCH often provides them with a good reference to a dog ethogram, especially an illustrated one for body language signals. Sometimes, JCH will even go on a walk with a client and practice "building" an ethogram together, thinking about individual behaviors, how to categorize them, how to define them in a unique and replicable way, and what we know about the role, or cause, of each behavior. This is a popular exercise, and he has been hired simply to provide this lesson to friends and families of his clients who get first-time dogs. He has delivered a similar workshop, "Let's Build an Ethogram!" for clients working with nonhuman primates in captive and wild settings, as well. I've even used the concept, for similar reasons, for parrot owners.

Cataloguing and describing an animal's behaviors is one of the most reliable and effective ways to get a glimpse into another species' possible emotional states, thoughts, perspectives, and intentions. For example, a dog mounting another dog could be playful, sexual, or aggressive, depending upon the corresponding cues given by both dogs. A "growl," defined as a guttural noise emitted from the throat with teeth together, could be aggressive or playful, given the behavioral context, and could also mark the shift from one context to another, such as when play becomes too rough and shifts into agonistic behavior.[66]

Let's revisit the concept of our dogs feeling "guilty," but through the umwelt of our dogs. How do dogs perceive the consequences of their actions in social settings? Play is an activity that's demonstrated throughout the animal kingdom, and it's one context that animal behavior researchers have analyzed to assess how a species might demonstrate consequences for their actions – and how they respond to these consequences. Research on dog play conducted by ethologist Dr. Marc Bekoff is shedding some light on whether dogs have their own version of "morality" or not, including how they react to perceived "transgressions."

Play is an important behavioral context for research because it's exhibited by a large range of animal species, which appear to have species- and group-specific gestures and "rules" that players follow. Watch a group of dogs at a dog park and you'll see an intricate interplay of postures, vocalizations, and movements. One dog might invite another to play with a "play bow," a specific gesture that entails a raised rump, with his front paws outstretched before him. The recipient dog may then reciprocate this gesture, and play, including dog laughs, barks, a bouncy gait, and wagging tails, ensues. There's nothing random about this series of behaviors. According to Bekoff,

---

[66] See image for full ethogram.

these play behaviors hint at something similar to humanlike morality, and during their play, he states that dogs exhibit honesty and deceit, invitation and clarification, warning and apology. For dogs, a "play bow" is an invitation to play, but it's also insight into canine morality. A play bow is a universally used and recognizable dog gesture, just as all chimpanzees, captive or free-living, will raise a leg behind them toward a recipient to invite them to play. According to Bekoff, these are specific, meaning-laden gesture, and there are moral and social codes to responding to them. One gesture might mean, "I'm biting you, but it's all in play," while another might be an "apology" when play becomes too rough.

Bekoff analyzed video footage of dogs, wolves, and coyotes across multiple behavioral contexts, including play. During his observations, he noticed and wondered about play, which can be calorically expensive and potentially make an individual susceptible to injury or predation risk. The canines' play initially appeared to have no other purpose other than the fact that it was pleasurable, but then Bekoff looked closer. In addition to behaviors like grooming and food sharing, play is an important behavioral context for clarifying and cementing social relationships. Gestures like the play bow hint at a canine's understanding of the consequences of their actions. During his research, Bekoff found that dogs would self-handicap to enable continuation of play. A large dog, for example, might change her body positioning or the tempo of her play to allow a smaller dog to keep up. Similarly, an adult dog might be more tolerant when playing with a very young dog than they would be when playing with a same-age peer. Dogs appeared to understand that their actions needed to be appropriately edited, and responded with "apologies" when they had transgressed by playing too rough.

Bows, nips, squints, yelps: these are all signals that are important during play to ensure that when the interactions become high-arousal, the behavioral context doesn't change to agonistic. Appropriately responding to these signals during play aren't just important to domesticated dogs; they're important to their wild cousins, too. Free-living coyotes will exclude those who don't respond appropriately; for example, if the animal bites too hard or pushes another animal over. The other coyotes will take note this failure to follow the rules and will cease playing with the rule-breaking canine. Coyotes who have been reprimanded for "unfair" play will slink away with their tails down, heads cocked to the side, and ears placed slightly back. The reprimanded coyotes then exhibited approach-avoidance behavior. According to Bekoff, it appeared as though the coyotes were apologetic and were seeking forgiveness. So while your dog doesn't feel "guilty" – at least not the way that humans do – he does have his own sense of morality.

So what happens to the communication between two dogs when one or both of them has had their ears cropped or their tail docked? In addition to ethical concerns that have been raised about these procedures, it could be argued that altering the natural conformation of the dog in these ways can potentially be communicatively handicapping. When a tail wag and ear placement, in conjunction with a suite of other behaviors, can mean the difference between friendly and unfriendly advances, the absence of these signals can create confusion and inappropriate responses.

From observing behaviors, we can hypothesize that an animal might have an understanding of another individual's perspective. For example, a chimpanzee might use visual gestures only when an individual is oriented toward them, reserving tactile and auditory gestures for inattentive recipients. This would indicate that the chimpanzee might understand that a recipient can only effectively receive a visual gesture if their face is oriented toward the gesturer.[67] We can also find insight into an animal's assessment of another individual – from their body posture, movement, and facial expressions, do they look relaxed or uneasy, submissive or aggressive? A dog might yawn or lick their lips when they're nervous; make a "play bow" when soliciting play; or sneer and growl when feeling threatened. Each of these behaviors would be fastidiously catalogued and coded. While dogs cannot use spoken language to tell us their thoughts, feelings, and intentions, humans have become adept at reading their body language – almost as adept as dogs have become at reading ours.

Ziva's behaviors – including the placement of her tail, and how it moved; how high her head was raised, how her nostrils moved and how her ears rotated; and the placement of her body – these would all be part of an ethogram. To a naive observer, these behaviors might be difficult to understand without a diagram that spelled out the meaning of each placement, configuration, and movement. For Alec, however, when Ziva looked up at him, stood stiffly with her tail raised, or put her ears back, these were all behaviors rich with important information. Alec could read these actions, just as his ancestors had, for generations before him. "Begging" behaviors from dogs likely originated when the human-canine relationship first began – and the first canine decided that the benefits of being with man outweighed the risks. The concept of an ethogram was likely born from man's relationships with other species and his attempts to understand them, even before he could have conceived of an umwelt, or that the umwelt of a dog differs so vastly from the umwelt of a human.

The key to a successful relationship with our dogs is fluency in both canine communicative signs and in their unique sensory perspectives. What might be salient to one species (temperature or butyric acid for the tick; a high reliance on visual cues and the propensity to create facial patterns for humans; olfactory cues for domesticated dogs) won't necessarily be salient for another.

WHY DOES MY DOG DO THAT?

Even though we have shared our lives with dogs for millennia, they are still apt to behave in ways that mystify us. Watch a dog as he or she finds a place to relax. They will likely inspect the area, rotate several times, perhaps pause and paw the area with a forepaw, rotate again, and then finally lie down. But why do they do this? Spinning several times before curling up appears to be a hard-wired behavior from the time

---

[67] Campion, T. L., Jensvold, M. L., & Larsen, G. (2011). *Use of gesture sequences in free-living chimpanzees (Pan troglodytes schweinfurthii) in Gombe National Park*. Tanzania. http://onlinelibrary.wiley.com/doi/10.1002/ajp.20978/epdf.

when dogs would have to make a nest before they settled in for the night. Most of our dogs, excepting dogs who live and work primarily outdoors, no longer need to make a "nest"; they might need to rearrange the dog bed or the blankets, but that's usually the utilitarian extent of this behavior.

Watch as a dog reclines on the couch with you or greets you after a long absence. They might want more physical contact and they might even lick you, too. But why do dogs lick humans?

Licking is a species-specific behavior among dogs, beginning from a very early age. Young dogs will lick the mouths of older dogs to encourage them to regurgitate what they've just eaten (sharing is caring!), as it can be harder for younger dogs to obtain their own food. But licking behaviors persist beyond puppyhood and cross the species barrier. With humans and dogs, licking is equal parts affection (even Darwin noted that this appeared to be an affectionate gesture), appreciation of our salty epidermis, and habit; some dogs might also lick their people for reassurance or attention.

Observe a dog who is panting heavily. Taken separately, and without a suite of other behavioral indicators, panting can be indicative of many different physiological states. Not only is this how they "sweat," but dogs also pant when they're warm or when they're in distress. Jack, for example, will pant when he's feeling nauseous; panting and licking his lips are two possible indicators of gastrointestinal distress.

Recall a time when you were sitting on the couch with friends or family. Why would your dog come up to you and sit on you – or wedge themselves in between you and another person?

While some people believe that these behaviors are indicative of dominance, it can also simply be affection, reassurance because they're feeling insecure, or simply the enjoyment of contact with their humans. Our dogs enjoy being the center of attention – what better way to ensure this than by inserting themselves into the center of the social interaction?

Try speaking to your dog in an inquiring tone. Does he or she tilt their head from side to side as you speak? Dogs will tilt their heads to adjust their outer ears (pinnae) while focusing on the origin of a sound. The behavior is both attractive to the human recipient and utilitarian for the dog who is taking in the auditory information.

Many dogs chase their tails, but why? Tail chasing can begin with one of several reasons, and if it goes unchecked, it can move into the territory of stereotypical behaviors. Dogs might chase their tails out of boredom, because they have irritated anal glands, because they have fleas or dermatitis and their hindquarters are itchy, because their movement has been restricted in a small space, because of hereditary tendencies, particularly in bull terriers, Australian cattle dogs, and German shepherds, and because of canine compulsive disorder, which can be treated with Prozac. Dogs will also chase their tails during play; puppies will chase their tails when they're first learning that this appendage is actually part of their bodies.

We've all heard dogs howling – it's a disquieting sound for humans. Dogs can howl for a number of reasons, but this typically indicates anxiety or an attempt to locate others (like a homing beacon); howling can also be indicative of pain resulting from illness or injury.

If you've ever scratched a dog's underside, you've likely seen that one or two-legged bicycle response. But why do dogs kick their legs when you scratch their stomach? Does it really feel so good that they can't help themselves? Actually, it's possible that they're kicking out like that because it tickles them or it's irritating to them, as well.

If you've ever been on a walk with a dog, you've probably seen a suite of "unusual" behaviors, such as the flehmen response, where a wide range of animals, including dogs, cats, and horses, curl their upper lip while investigating a smell. Physiologically, the flehmen response facilitates the transfer of odors into the Jacobson's organ, which is located above the roof of the mouth, but physically, the curled back upper lip and exposed teeth look completely bizarre. Anyone who has shared their life with a cat, dog, or horse has probably seen this behavior, especially after the animal has smelled something pungent, like urine or feces, or after they've ingested something unpleasant, like a horse who has just received worming medication. Other strange behaviors include eating grass and rolling in smelly areas that they've found. Not all dogs like to roll in smelly objects, but some of them can't seem to get enough of it. So why do they do this, particularly when their sense of smell is so much better than ours? Isn't the odor overwhelming to them? It's possible that dogs cover themselves in smelly substances to "show off" to their peers – human and canine alike. In the dog world, strong scents are highly esteemed, so being smelly is a "good" thing for them. Odors that turn our stomachs might smell positively divine to your dog. We have to remember to consider our canines' perspectives; humans prefer scents that are "clean," fragrant, and fresh, while our dogs prefer scents that, to our perspectives, are rank, rancid, and repulsive.

Why do dogs lick or consume urine, participate in coprophagy (eating feces) and participate in other strange behaviors? Dogs' bladders have the capacity to release numerous small amounts of urine, allowing them to mark their territory many times over the course of one outing. When he goes for a walk, it's not uncommon for Jack to urinate two dozen times or more. Urine is a scent marker, so dogs urinate, in very small amounts, to leave a "calling card," as it were. The dogs that follow along read that card by sniffing thoroughly and even licking the location. Coprophagy, on the other hand, has been more of a mystery to animal behaviorists. We seem to be unraveling the mystery of "poop" eating recently: we have discovered that in some species, critical nutrients are produced by bacteria in the large intestine, but cannot be absorbed there. Hence, to acquire those critical nutrients, you (not you specifically, but your dog), eat your own feces. This is a common behavior in some herbivores, like rabbits and some rodents, but not so common in carnivores like wolves and dogs. In my (JCH) experience, coprophagy in dogs tends to be a sign of nutritional deficiencies or a metabolic-based disease, and needs the attention of a qualified veterinarian.

Why do (many) dogs love to eat cheese? Cheese has a protein called casein, and when it's broken down in the stomach, it produces a peptide called casomorphins. These are opiates that act as histamine releasers (which is why as many as 70% of people worldwide have some level of lactose intolerance). The opioid casomorphins also trigger the reward center of the brain, though, just like a drug would. Cheese acts on the brain in a way similar to morphine and heroin, so it's no wonder that our dogs (and humans, too!) are crazy about it.

Another odd behavior for a carnivore like dogs is chronic grass-eating: again, it was a mystery for many years, and recent research from the University of California-Davis has shown a relationship between grass-eating and gastrointestinal issues, especially parasite infection. When JCH notes chronic grass-eating in a dog, he immediately recommends a very thorough veterinarian check-up.

All of these behaviors are signals; your dog is communicating information to you, and in order to truly understand your dog's behavior, it's critical to understand these signals correctly. A tremendous proportion of clinical dog behavior problem cases arise because of a fundamental misunderstanding in what your dog is saying. In fact, by mis-interpreting your dog's signals, you can make a minor situation much worse. As we will talk about later, some breeds are better at communicating, and many simply communicate differently: it's a breed-specific skill to be able to accurately communicate with your dog, or your friend's dog.

FOUND IN TRANSLATION

So how do our dogs communicate with us and with each other? The ethogram is an essential point of departure for understanding canine communicative modalities, which consist of olfactory cues, vocalizations, facial expressions, eye gaze, and gestures, including body posture. A dog's auditory communication, which includes barks, growls, howls, whines, whimpers, screams, pants, and sighs, are context-specific and vary in pitch, amplitude, and timing that can potentially alter the meaning of the vocalization.

Barks comprise only 2.3% of all vocal communication with wolves, whose vocal repertoire includes eleven "sound types" identified by sonographic analysis.[68] Dogs bark far more than wolves do, and across all behavioral contexts.[69] Spectrographic analyses reveal that the structure of dogs' barks varies with these contexts,[70] too. There are many different kinds of bark, including rapid, mid-range pitched "alert" barks, as in the bark that the new dog gave when he sensed Alec and Ziva; barks that rise in pitch, indicating play, stutter barks to initiate a play session, brief barks when startled, lower pitched, slower barking to indicate a threat, one or two short barks can indicate annoyance (in lower pitches) or greeting (in higher pitches), and a series of barks with intervals between them can indicate a dog who needs companionship. Unlike wolves, African wild dogs are also very vocally inclined and use a wide range of calls, including barks, yelps, growls, squeals, cries, whimpers, and whines.

---

[68] Schassburger, R. M. (1987). Wolf vocalization: An integrated model of structure, motivation and ontogeny. In H. Frank (Ed.), *Perspectives in vertebrate science, Vol. 4. Man and wolf: Advances, issues, and problems in captive wolf research* (pp. 313–347). Dordrecht, Netherlands: Dr W Junk Publishers.
[69] Coppinger, R., & Feinstein, M. (1991). Hark! hark! the dogs do bark. and bark and bark. *Smithsonian, 21*, 119–129.
[70] Yin, S. (2002). A new perspective on barking in dogs (*Canis familiaris*). *Journal of Comparative Psychology, 116*(2), 189–193.

It's possible that barking was selected for, intentionally or unintentionally, early on during the domestication process. Today, many modern dog breeds have a propensity to bark (such as German Shepherds and other breeds that have traditionally been bred for guarding or herding) or to be quiet (such as hunting dogs that rely upon stealth.) Over more than 30 generations, Belyaev and Sorokina's foxes, in addition to becoming more docile, having floppy ears, and demonstrating the piebald coloration patterns associated with domesticated animals, also exhibited more of a propensity to bark (or "yip") than their wild relatives did, especially during the "greeting" context with humans.

The vocal communication of dogs isn't limited to barking, though: they also howl, growl, scream, pant, sigh, whimper, and whine; their auditory repertoires are vast and varied. Growls can have a low or midrange pitch, and typically indicate a threat, but many dogs also use "play growls" during the play context. Howls are mid-to-high-pitched vocalizations that typically indicate loneliness or the presence of other canines. Screams are sharp, high-pitched vocalizations that can be alarming for human recipients to hear, and rightly so: they typically indicate agony or a call for help. Pants are typically demonstrated when a dog is trying to regulate their body temperature or when they are in distress or discomfort, but they also occur when a dog is happy, excited, or during the play context (breathy pants are how both dogs and chimpanzees laugh). When Ziva was walking with Alec, she panted quickly with a relaxed jaw; she was happy after having received treats from him and excited that they were approaching the blind. Later, when she was inviting the new dog to play, she emitted a series of breathy pants. Sighs can indicate contentment when the eyes are closed, attention-seeking when attempting to have an interaction, and displeasure when the eyes are wide open. When Ziva was walking with Alec, she made a "whuff" sigh to him to get his attention. Whimpers and whines can indicate pain, fear, pleasure, or pacifying behaviors, depending upon the pitch and length of the whimper or whine.

Many of the above behaviors referenced "calming signals" or "self-calming behaviors." According to Turid Rugass, author of *On Talking Terms With Dogs*, there are approximately 30 "calming signals" that dogs use to alleviate tension in their social interactions with other dogs and with humans. These signals include, but are not limited to: yawning, "smiling," licking, turning the head away/averting gaze, sniffing the ground, play bow, freezing, sitting down, walking slowly, walking in a curve, wagging the tail, lying down, and urinating. Humans also use calming signals, including yawns, to diminish stress.

Dogs, like humans, communicate in multiple modalities simultaneously, using distance (proxemics), body language (kinesics), touch (haptics), the involvement of the eyes, including blink rate, gaze holding, and pupil dilation (oculesics), and nuanced meaning in their verbal communication, including pitch, prosody, intonation, and volume (paralanguage).

All of these behaviors, when taken into account with vocalizations and other behaviors, can paint a complete picture of canine communication. But what's beyond what we can observe?

A dog's three-dimensional olfactory world and their range of hearing and vision play important roles in their training. Even though we can't sense smells the way that our dogs do, we can observe how they process them. Scent dogs can smell a track that was laid hours before, and can be effectively trained to follow this smell; this is something that no human child has been able to effectively learn. Conversely, human children can be taught to write, but this is something that no domesticated dog has been able to effectively learn. The objective has to be realistic, given the species' umwelt. While we can't see, smell, hear, or feel the world from a dog's perspective, we can use tools such as ethograms and the results of physiological research to better understand how they navigate the world – and thus how we can be effective and efficient teachers for them. Understanding the species- and breed-level motivations and umwelts of each dog can help dog owners enable their dogs to successfully learn – and enable their owners to partner with them more effectively. And while popular culture likes to say that dogs "feel guilty" about human-oriented "transgressions," there's little scientific evidence to support this: that look to which we're so quick to assign "guilt" is actually an altogether different emotion, translated through our own human lenses. That supposedly "guilty" look on our dogs' faces says just as much about us – and our history of reactions to our dogs' actions – as it does about our dogs' anticipation of how we're going to react.

AN EXAMPLE OF AN ADULT, NON-REPRODUCTIVE DOG ETHOGRAM (WITHOUT SEXUAL AND INFANT CARE BEHAVIORS)

Low Activity	
Lying	torso on the ground
Sitting	rear legs on the ground, front legs extended
Standing	torso off the ground
Moderate Activity	
Exploring	low level locomotion, sniffing, and visual investigation
Social	low level, casual investigation of other dogs and play
High Intensity Activity	
Trot	lower speed locomotion
Gallop	intermediate speed travel
Run	full-speed travel
High Confidence	
Tail up	self-explanatory
Ear up	ear fully erect and displayed
Moderate Confidence	
Neutral ears	ears to the side
Neutral tail	tail at or below horizontal
Low Confidence	
Ears back	ears flat to scruff/neck
Tailed tucked	self-explanatory
Low Intensity Social	
Greeting	approach to another dog with intent to interact
Nose touch	self-explanatory
Inguinal sniff	sniff of the groin area

Moderate Intensity Social	
Rally	multiple dog participation/recruitment
High Intensity Social	
Play bow	front legs stretched forward, front body lowered
Stalking	stiff-legged gait, tracking and following another dog
Play biting	biting without intent to break skin, grasping only
Flight	
Escape	self-explanatory
Leave	casual, low-intensity departure from social situation
Mixed Fight and Flight	
Whirling	spinning movement to escape attack
Defense snap	an "air snap" used as a signal of arousal
Snarling	characteristic vocalization in mixed state behavior
Fight	
Lunge	self-explanatory, with intent to bite with damage
Wrestle-fight	self-explanatory, with intent to bite

Adapted from work by Jane Packard and Sonia Alvarez, based on work by Roger Abrantes, 1997 (*Dog Language*) and Erik Zimen, 1975 (*The Wolf*).

AN EXAMPLE OF A SIMPLE DOG ETHOGRAM

For use in a shelter-based research project (i.e. for singly-kenneled dogs) by J.C. Ha.

Movement
   Pacing
   Jumping
   Hopping
   Walking

Postures
   Laying
   Sitting
   Stand: all fours
   Stand: rear legs

Location in kennel
   Front
   Middle
   Back

CHAPTER 3

Mapping the mind: Brain structure and development

MAPPING THE MIND

How do you map the world – or the mind – without a clear understanding of what you're documenting? This is the conundrum that early cartographers, psychologists, and later, neuroscientists, faced. Neuroscience, like cartography, has had its fair share of historical "wrong turns." During ancient times, it was commonly believed that the Earth was a flat disc, and while the concept of a spherical Earth was proposed by many, including Pythagoras in the sixth century BCE and Aristotle in the fourth century BCE, this didn't stop many early cartographers from documenting the Earth as a smooth surface that lost ships would tumble over. Many early maps depicted the planet with landmasses or seas that were absent or misplaced – or even accompanied by mythical beings. Maps of the "Lost City of Atlantis" have existed for centuries and early maps of the "New World" involved equal parts exploration, anecdote, and guesswork – California was once depicted as an island. Olaus Magnus' *Carta marina* (map of the sea) of 1572 contains numerous mythical sea monsters, including a giant sea serpent attacking a ship in the sea near Norway and a creature known as a "ziphius" eating a seal while another monster attacked it.[71] It was only after extensive exploration and countless missteps along the way that we gained enough knowledge about the shape and structure of our planet to map it accurately; maps of a round world finally began to emerge during the Hellenistic age, which began in 323 BCE.

The brain proved to be as enigmatic as the mythical monsters of the *Carta marina*. Early civilizations struggled to understand the brain's structure and function even more so than they struggled to map the planet. While tenacious travelers could reach the far ends of the earth, how could early scientists determine how the brain worked, especially when some civilizations had prohibitions against dissection? And even when these prohibitions didn't exist, misperceptions still did. Ancient Egyptians believed that the heart was the body's cognitive epicenter; Aristotle thought that the brain's function was to provide thermoregulation for the heart; and the pseudoscience of phrenology used skull measurements to determine an individual's personality and proclivities. While the first attempts to map the mind were external and yielded only pseudoscience, later scientists discovered that there are internal parts of the brain that do correspond to certain functions, including personality. The classic example proffered

---

[71] http://www.utaot.com/2013/09/09/mythical-creatures-as-drawn-on-medieval-maps.

64 CHAPTER 3 Mapping the mind: Brain structure and development

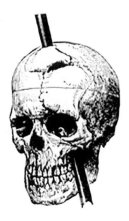

FIG. 20

Phineas Gage with the rod that pierced his skull and a depiction of how the rod entered and exited his head.

by introductory psychology textbooks is that of Phineas Gage, an American railroad worker who sustained a catastrophic injury to the left frontal lobe of his brain when an iron rod was driven completely through his skull. Gage's left frontal lobe was almost entirely destroyed – as was his prior personality: he reportedly lost his inhibitions permanently and exhibited violent or inappropriate tendencies that he hadn't demonstrated before (Fig. 20).

Gage's gruesome accident was likely the first case that provided evidence that damaging certain parts of the brain could lead to changes in personality. Advances with magnetic resonance imaging (MRI) have yielded equally surprising results about how both human and canine brains work – and just how much we have in common. But how can we map the brain in practical terms to understand the similarities and dissimilarities between human and canine brains?

IF I ONLY HAD A ~~BRAIN~~ NEOCORTEX

The mammalian brain, which is the body's most complex organ, has some very simple and humble origins. Brains, which consist of groups of nerve cells called neurons, first emerged in living beings approximately four billion years ago, with our earliest single-celled ancestors. Externally, the brain looks a lot like a giant walnut, but internally, it's like a thriving city transected with neurons that act just like highways. Neurons are the fundamental part of all mammalian brains, and they include two main types of brain tissue: white matter and gray matter. White matter consists primarily of axons, while gray matter is comprised of synapses. An axon is the threadlike part of a nerve cell that conducts electrical impulses from the neuron's cell body. Neurons are arteries for information to spark between one location to the next, whether you're a reptile, a bird, a fish, a canine, or a human. And synapses are the junction between two nerve cells, demarcated by a gap across which the impulse passes.

The brain is comprised of three main parts: the forebrain, the midbrain, and the hindbrain. The forebrain includes the cerebrum, hypothalamus, and thalamus. The midbrain includes the tegmentum and the tectum. And the hindbrain contains the cerebellum, pons, and medulla. The functional parts of the brain include the Motor Area, which controls the voluntary muscles; the Sensory Area, which corresponds to skin sensations; the Frontal Lobe, which corresponds to behavior, personality, moods, movement, problem solving, and concentration; Broca's Area, which corresponds to speech control; the Temporal Lobe, which corresponds to memory, language, and hearing; the Parietal Lobe, which corresponds to language, perception, body awareness, sensations, and attention; the Occipital Lobe, which corresponds to perception and vision; Wernicke's Area, which corresponds to language comprehension; the Cerebellum, which corresponds to coordination, balance, and posture; and the Brain Stem, which corresponds to heart rate, breathing, and consciousness[72] (Fig. 21).

The brain, simply put, is an animal's master control center. It's a part of the central nervous system (CNS), a component of the nervous system that integrates and processes information and coordinates responses. Among mammals, the central nervous system comprises the brain; spinal cord; peripheral nervous system (PNS), which includes the nerves within the body; and the neocortex, the outermost portion of the cerebral cortex. Of all extant vertebrates today, mammals alone possess the neocortex, which is highly developed among humans in comparison to other mammals.

On average, the human brain weighs approximately 2.7 pounds, in comparison to an elephant's brain, which weighs 11 pounds, a cat's brain, which weighs approximately 30 grams (.066 pounds), and a dog's brain, which weighs approximately 72 grams (.16 pounds). In comparison, a gray wolf's brain weighs approximately 119 grams (.26 pounds). The average dog's brain weighs less than one-half of one percent of their total body weight, yet it receives more than 20% of the blood supply pumped from the heart. But it's not just the size of the brain that matters – it's the encephalization quotient, or EQ, which measures relative brain size as the ratio between the actual brain mass and the brain mass that would be predicted for an animal of that size. Encephalization is thought to be predictive of intelligence; animals that have higher EQs are said to have higher cognitive capacities, while animals with

---

[72] See Labelled brain map.

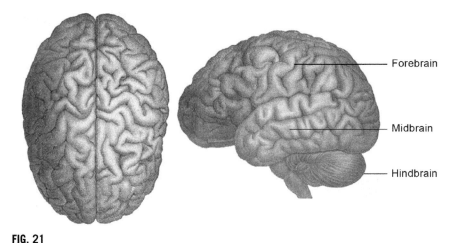

FIG. 21

The human brain. Notice that there is less "folding" in the canine brain (Fig. 22) in comparison.

lower EQs are said to have comparatively lower cognitive capacities. Humans have an EQ of between 7.4 and 7.8; elephants have an EQ of 1.13 to 2.36; dogs have an EQ of approximately 1.2; and cats have an EQ of 1. Also of importance is the number of neurons – those superhighways that convey information – in an individual's brain. In the cerebral cortex, where higher cognitive functions reside, humans have 19 to 23 billion neurons; African elephants have 11 billion; cats have 300 million; and dogs have 160 million. And despite these numerical differences – and millions of years of evolutionary separation (dogs and humans last shared a common ancestor as long as 100 million years ago), the two species have striking similarities in their brain structure and function.

The mammalian brain differs from that of other animal species, such as reptiles, fish, and birds. The reptilian brain, which includes the brainstem and cerebellum, is the most primitive of the three evolutionary portions of the brain. The reptile brain forms the upper portion of the spinal cord and is responsible for regulating vital functions such as respiration, digestion, thermoregulation, heart rate, our vestibular system, and blood pressure. The limbic or mammalian brain emerged among the first mammals and provides the capacity for consciousness and emotions. The limbic brain includes the hypothalamus, amygdala, and hippocampus. The neocortex, which is the most recent evolutionary portion of the brain, is most developed among humans and is comprised of two cerebral hemispheres that play a role in abstract thought, consciousness, and language.

The largest portion of the brain is the cerebrum, which is divided into two halves called cerebral hemispheres. These cerebral hemispheres are further divided into four lobes: the frontal, occipital, parietal, and temporal. The frontal lobe is the brain's emotional control center and is integral to forming our personality, concentration, problem solving, planning, body movement, emotional reactions, speech, and smell. The frontal lobe comprises 29% of the brain for humans, 7% of the brain for dogs, and 3.5% of the brain for cats.[73] A dog's world is chiefly navigated by olfaction and approximately one-third of the canine brain is devoted to olfaction. The olfactory portion of a dog's brain is 40 times greater than it is among their human companions, so it's no wonder that dogs experience the world through "smellivision," as Dr. Temple Grandin calls it.

Consider the concept of a homunculus, a term that originally referred to a fully formed (albeit miniature) human and which first appeared in the writings of Paracelsus (1493-1541) in his *De natura rerum*. In psychology, a homunculus refers to the sensory map of a species' brain, depicted in terms of their body parts. For example, the homunculus of a dog's brain would depict a body that had a very large nose and ears, while the homunculus of a human's brain would depict a body that had an oversized tongue and fingers (Fig. 22).

The parietal lobe controls body awareness, touch and pressure, and taste. The temporal lobe controls facial recognition, auditory processing, and long-term memory. And the occipital lobe is chiefly responsible for vision. The cerebellum, which is Latin for "little brain," control balance, coordination, and fine motor movements; and the limbic system controls emotions such as sadness, happiness, and love. To understand how our dogs process information, it's important to understand their brain structure and development. This begins with a comparison between dogs and wolves,

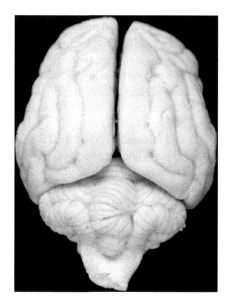

FIG. 22

Canine brain image.

---

[73] Fuster, J. M. (1997). *The prefrontal cortex*. Philadelphia: Lippincott-Raven.

who are their closest evolutionary relation, and dogs and humans, who have co-evolved for millennia.

Gray wolves and domesticated dogs share most of their DNA in common, but within that .4% of difference, there are numerous divergences in gene expression that are demonstrated in neurological and behavioral dissimilarities. After generations of co-evolution with humans, the brain of the domesticated dog demonstrates qualitative differences from the brain of wolves. In a 2004 study,[74] researchers found that the hypothalamus was highly conserved among the wild canids, while the hypothalamus of domesticated dogs had a high proportion of gene differentiation. This is pertinent because the hypothalamus links emotional, endocrinological, and autonomic responses to exploratory behavior. The researchers wrote that this "suggests that the domestication process of dogs has greatly accelerated the rate of divergence in gene expression in the hypothalamus."[75] In addition to neurological distinctions, dogs and wolves also differ in their behavior and in the rate of their development, which has important implications for the human-canine relationship. Comparisons between the brains of dogs and humans reveals that while they are evolutionarily separated by tens of millions of years, there are many sounds and smells that are salient to both species.

THE NOT-SO-TERRIBLE TWOS

Dogs, like humans, experience developmental milestones. While toddlers are said to experience the "terrible twos," that age isn't so "terrible" for our canine companions. Age two is an important developmental milestone for domesticated dogs, as this marks the beginning of inhibition and their ability to successfully perform in professional capacities. Many service dogs and military and police K9s aren't considered to be "fully trained" until they reach the age of two because they don't have inhibition yet. Ziva, who is not yet two years old, is learning to listen to Alec, but she still possesses the impulsivity of an adolescent dog. She tried to take more food than Alec offered and tried to initiate play with an unreceptive dog. As her development progresses, though, she will exhibit increased impulse control, judgment, and an ability to better gauge social situations.

Canines experience several important developmental phases. During their neonatal period, which occurs during their first two weeks of life, puppies are unable to hear, see, or control their own thermoregulation – they depend completely upon their mothers. During their transition period, which occurs from days 13 to 21, puppies begin to smell and taste; they also start to hear and see their worlds, as their ears and eyes begin to open. During the awareness period, which occurs on the following three days, puppies are fully alert and aware of their environments. Their

---

[74] Saetre, P., Lindberg, J., Leonard, J. A., Olsson, K., Pettersson, U., Ellegrena, H., et al. (2004). From wild wolf to domestic dog: Gene expression changes in the brain. *Molecular Brain Research, 126,* 198–206.
[75] Ibid.

brainwaves exhibit qualitative changes and they begin to actively learn. During their socialization periods, which last from three to seven weeks, puppies learn species-specific behaviors, such as body postures, vocalizations, and appropriate behaviors for different contexts, such as play, that help them socialize with other dogs. Puppies who leave the litter before seven weeks of age can experience difficulty getting along with other dogs, as these early experiences teach them what to do as much as they teach them what "not" to do. From seven to 12 weeks of age, puppies experience a "human socialization period." This is when they begin to bond with their human family members and it's also the phase during which learning happens at a rapid rate.

While wolves and dogs are both social canines who share 99.6% of their DNA in common, they differ in their social structures, behavior, diversity, and, perhaps most importantly, in their developmental trajectories. Across all developmental milestones, wolves are more precocious than dogs: wolves' eyes open earlier, they hear earlier, they walk sooner, and they begin to manipulate objects with their mouths earlier than their domesticated dog relatives. Wolves are more precocious than dogs are, and all of these milestones occur at comparatively earlier times than dogs. While there are some breed differences, developmental milestones for all dogs happen comparatively later than they do for wolves. This provides dogs with an extended window of opportunity to socialize with humans – and perhaps the first ancestors of modern dogs exhibited these slight delays in comparison to other canines. This slower rate of development would have been a definite advantage to the burgeoning human-canine relationship.

It's during the 18-24 month stage that the relatively slower rate of development might call upon our patience the most. It's a good thing that dogs, like many domesticated animals, retain highly attractive neotenous[76] characteristics and bond so completely with us – it helps us deal with the behavioral issues that are likely forthcoming. From 18 to 24 months of age, puppies experience young adulthood; this is akin to the human teenage years, and for dog owners and trainers alike, patience and a sense of humor are crucial for success. Think of your juvenile dog in terms of a teenager who's venturing out into the world: they're going to have the outward appearance of a sub-adult, almost-grown individual, but they're going to lack the judgment and physiology to consistently make "good" choices, just as Ziva struggled to make. Ziva wanted to make Alec happy, but she also wanted another piece of jerky; she chose to try the latter instead of patiently waiting for another piece. Human teenagers often partake in high-risk behaviors such as staying out late, drinking, taking drugs, and exceeding the speed limit. "Teenage" dogs like Ziva also have impulse control issues that make it difficult for them to realize their full potential, whether as a pet or as a dog who has a particular job. There's a reason that *Doogie Howser, M.D.* was a popular, novel show during its time – while an adolescent might have the intelligence and the promise to one day be a successful physician, they typically still lack the

---

[76] Traits typically seen only in juveniles.

inhibition and judgment that an adult aged 25 or older would have. Similarly, service dogs and K9s typically don't officially begin their professional careers until they are two years old or older.

Lab mix Jack arrived at his new home at approximately 18 months of age. He knew the basics, but hadn't yet mastered concepts like "wait." He was enrolled in obedience training, and was doing well in the class, when the trainer decided one day to use him as the "example" dog. Jack, who was normally quiet, would whine when he needed to urinate; when he was asked to do a command that he didn't understand, rather than wait to go outside to go to the bathroom, he promptly lifted a leg on the trainer and emptied out the entire contents of his bladder, much to TLC's chagrin. Jack didn't return to that class, but his training did continue at home, where he started to learn how to "wait" – whether it was waiting for a treat or waiting to go to the bathroom.

JCH has seen hundreds of in-home cases of misbehaving dogs and owners – and he's documented developmental trends in this behavior, particularly when a dog reaches two years of age. Consistently, in the cases of the development of aggression and anxiety, when he asks them when the behavior began and encourages the owners to think back carefully, often years earlier, we can pin down the start of the problem to a dog that's one and-a-half to two years old. He credits this phenomenon to a *sensitive period* around this age in most breeds in which two changes in the brain occur: the final development of inhibition, and the process of social awareness, where a dog is introduced to the social adult world. He equates this to humans – or male humans, at least – at about 18 to 22 years of age. They're starting to be treated as adults; they can vote, fight for our country, and at 21, they can legally drink alcohol; yet sometimes, they're still treated as children, and lacking full and complete inhibition until 22 years of age. It's a time of social confusion and of recklessness, and the same sort of process occurs in dogs.

One of his cases was particularly troubling and involved an Australian Shepherd who had been acting aggressively towards children. Riley was about three years old, and sure enough, the behavior had started when he was about two. When questioned further, the owners admitted that Riley was a bit of a replacement for empty-nest children who had moved out. They stated that they'd never had a dog before, and that, when Riley began to 'act out' at about two, they simply isolated him, keeping him out of any social situations from which he could learn to behave appropriately. Riley had never had any significant training, although he had been taken to several socialization-heavy puppy classes. He was their "baby," so there were very few rules, and what rules did exist were enforced only by the husband, and never by the wife, leading to further confusion for Riley.

JCH finds that the prognosis in these cases is generally very good, as long as there was sufficient earlier, positive human and dog interactions. Successful interventions include prescribing a program heavy on rules and structure for interactions, with repeated, limited, structured social interactions with positive reward. Basically, we needed to teach Riley how to interact with others by implementing guidance and rewarding appropriate behavior.

Australian Shepherds like Riley love to have jobs and they are typically quick learners. Riley, being an Aussie, took right to the program and the owners saw significant positive changes within days. He's now a very happy and socially he's a well-adjusted dog who was actually a wonderful "trainer" for a second dog that this family obtained later on.

Interestingly, domesticated dogs learn the concept of object permanence earlier than humans do at an equivalent time in their development. Object permanence is the understanding that when an object is no longer visible that it hasn't just ceased to exist, but rather, continues to exist but is not detectable. The difference in understanding this concept can be gleaned from babies throwing spoons or rattles from their high chair, and then not looking for them. For the baby, the object ceases to exist. Conversely, a human could throw a ball for a dog, and the ball could go out of sight, but the young dog would continue to search for it.

HOW DO DOGS LEARN?

Our dogs learn in the same way that every animal learns, with the added advantage of being able to understand the desires of their human cohabitants. But what is "learning?" At its most basic, learning is a cognitive function based on genetic, inherited structures in the brain. Fundamentally, it is the change in behavior based on experience. A little more specifically, it's a change in the probability of a certain response based on the consequences of prior expression of the same, or a similar, response. Operant conditioning is a type of learning that occurs based up on the associated rewards and punishments for behavior. If you perform a behavior, a response, and receive an aversive (punishing) response, your probability of performing that response, especially to that same trigger or stimulus, decreases. On the other hand, if you receive a positive response (in your perception), your probability of performing that behavior increases. This is punishment and positive reinforcement learning mechanisms in a nutshell. We should note that, to be symmetrical about it, 'punishment' is often referred to as positive punishment. Perform that behavior and punishment is applied: positive punishment, a converse to the application of a positive response, or reinforcer: positive reinforcement.

Slightly more complicated, and less common, are the other two forms of operant learning (there are only four forms): negative reinforcement and extinction (sometimes, symmetrically, referred to as negative punishment). Negative reinforcement, which is *commonly* confused with punishment, is the *removal* of a negative response: this causes an *increase* in the triggering behavior (hence the "reinforcement" part of the name). A classic example is the aggravating alarm which goes off if you do not latch your seatbelt after starting the car. When you do latch your buckle, the sound goes off. By latching your seatbelt, you can make the bad sound go away: you become more likely to more quickly latch that seatbelt. That's negative reinforcement: removal of an aversive response *increases* the rate of the triggering behavior.

Conversely, extinction or negative punishment is the removal of a positive response: if you are being rewarded (thus increasing, or at least maintaining, a rate of behavior), and it is removed, you are going to be *less* likely to perform that behavior again. Once you are no longer being rewarded for maintaining a certain response (behavior), you stop doing it! At least, as much or as easily as you did it before… These four mechanisms together are how we, and our dogs, and all animals, learn. There are more details, more complexity, and more forms, like associative versus classical conditioning, habituation, taste aversions, observational learning, but these four concepts underlie a significant part of dog training and learning.

SAVED BY THE BELL

Dogs, like humans, process and retain information when they're learning. Numerous factors, including one's environment, early experiences and socialization, and emotional and cognitive influences help shape how information is retained, and some stimuli are more salient for some species than they are for others. Conceptual frameworks called "learning theories" describe the process of information acquisition. Learning theories include, but aren't limited to, behaviorism, cognitivism, and constructivism. Behaviorism examines conditioning and rewards, cognitivism examines the individual learner and their cognitive capacities, including memory, and constructivism examines knowledge acquisition for each individual.

Early breakthroughs in canine research came from the behaviorist learning theory. Any student who has taken psychology 101 learned about Russian physiologist Dr. Ivan Pavlov's animal behavior research, which was groundbreaking in the early 1900s. But before his work with sounds, salivating dogs, and food rewards was groundbreaking, it was mostly messy and surprising. Pavlov, who was studying dogs' salivation response in relation to when they were fed, found that the dogs in his study would salivate even in the absence of food. Pavlov's lab assistant truly knew how to make an entrance: all he had to do was cross the threshold to the testing room and his dogs would begin to salivate. Seeing that the dogs anticipated the food reward simply with the presence of the assistant, Pavlov elaborated on his study. He later discovered that his dogs would make an association between an auditory or visual stimulus and food when the two were presented in close temporal proximity. Whether Pavlov's auditory stimulus was a bell, a tuning fork, a metronome, or whether it was one of many visual stimuli, the dogs learned that they were about to receive food. Pavlov referred to this form of learning as a "conditioned response" and this was later referred to as "Pavlovian conditioning."

Learning theory continues to be applied to canine research and to dog training. While dogs have an amazing capacity to learn, they need to have the right environment to do so. To be social and understand social cues from other dogs and from people, dogs need to learn early communicative skills. For Chaser the Border Collie, who learned more than 1,000 words, that extensive vocabulary was the result of the dedication of her owner, John Pilley. As a retired psychology professor, Pilley had the knowledge and time to devote to his dog's learning. With the same patience, dedication, and enthusiasm that he had used to teach his psychology students, Pilley taught Chaser for more than five hours per day, seven days per week, over the course of nine years[77] to eventually reach her 1,022-word vocabulary. Yes, Chaser learned, and she learned a lot, but she also had a teacher who was well-versed in human and canine behavior and who was uniquely capable of helping her reach her full potential. Just as Pavlov noticed that his dogs were salivating in the absence of the intended stimulus, Pilley noticed that the time needed to teach Chaser new words began to decrease. Not only was Chaser learning new words, but she was learning them at a faster rate, a pattern of learning that's also seen in human children. Her cognitive capacity, including her memory, was immense, and her individual knowledge acquisition, defined by her large vocabulary, remains a benchmark for other dogs.

THE NOT-SO-CURIOUS CASE OF GUS GARNER

Developmental psychologists describe sensitive and critical periods of development. Critical periods of development are times where the nervous system is particularly sensitive, and wherein something cannot be learned outside of that period. Sensitive periods are times when something can be learned rapidly and with more ease than when it is attempted outside of that period (for example, learning a second language.) For the human-canine relationship, critical and sensitive periods of development are particularly important. Dogs must have that early exposure and interaction with humans to cement the relationship. A dog named Gus was a beach dog who had lived his entire life on his own, honing skills like hunting and scavenging, but missing out on that early exposure. For dogs like Gus, a feral life on a beach in the Dominican Republic doesn't provide any early socialization and it can be difficult for them to build a relationship with a person later on. Gus' case isn't unique, though, nor is the Dominican Republic unique as a site for feral dogs. Feral village and beach dogs can be found worldwide, including in Africa, India, Israel, and the Americas, existing on the edges of human civilization. These feral dogs are tolerant of humans, but not necessarily trusting of them; their relationship is commensal, meaning that one species benefits from the other relationship, with no harm coming to the other, similarly to how some envision the beginning of the canine-human partnership. Most family dogs are raised in homes where they're familiarized with human sounds, smells, and sights, but feral dogs are peripheral and not central; they're aware of humans, but they lack the social skills of dogs that have been raised in the home. Rather than learning how to read and relate to humans from a young age, Gus was introduced to life in a human family when he was an adult dog. Gus was rescued and adopted by a kind person named Sundae Garner, but the two of them soon found that a lack of early exposure can have long-term effects (Fig. 23).

Because he had missed out on early socialization with humans, he was understandably standoffish with them, even in a nurturing setting where a typically socialized dog would have thrived. Numerous studies have found that dogs who have

---

[77] https://www.cbsnews.com/news/the-smartest-dog-in-the-world.

not been exposed to human socialization, especially early in their development, will demonstrate a tendency to withdraw from and not know how to interact with humans. Often, typical dog-human relationships cannot be established with dogs who have been deprived of early interactions with humans.[78] But demonstrating dysfunctional behavior after early social deprivation isn't limited to dogs; humans, too can exhibit aberrant behavior after early social deprivation. One of the most well-known cases of human social deprivation pertains to Romanian orphans who later exhibited atypical behavior similar to those diagnosed on the autistic spectrum and exhibited stereotypical behaviors, including rocking back and forth. Recent research demonstrates that psychological well-being and cognitive abilities correlate directly with the percentage of attention that infants receive when they are young.[79] One study found that human infants who are deprived of being held and spoken to can have comparatively smaller brain size, attention capacities, and emotional bonding skills than infants who are nurtured, held, and read to. Dogs, like humans, struggle to socialize when they're deprived during those formative months and years.

Again, to understand behavior, you have to integrate multiple levels of influence. More immediately, behavior is the result of established cognitive capacity, including emotional responses and personality dimensions, which process sensory perception as well as immediate environmental influences through learning and experience. Social interactions influence behavior at every level in this hierarchy, from evolutionary selection pressures to development of the individual to personality, and in an immediate sense as part of the animal's environmental interactions.

FIG. 23

Gus Garner.

[78] http://science.sciencemag.org/content/133/3457/1016.
[79] https://www.washingtonpost.com/local/romanian-orphans-subjected-to-deprivation-must-now-deal-with-disfunction/2014/01/30/a9dbea6c-5d13-11e3-be07-006c776266ed_story.html.

Dogs like Gus, who had already struck out on their own (and without human care and attention) already experienced those sensitive and critical periods. They were beyond their infancy; much like the Romanian orphans, they never received proper socialization. As a result, their developmental trajectories were altered. The prognosis for these dogs varies from those who can be helped with medication to those who are profoundly impacted – perhaps indefinitely. So why can the relationship between humans and dogs be so transformative?

The experiences and the environment in which a dog exists during the sensitive periods in the development of its brain are critically important to the development of its later behavior. These early experiences are a crucial part of mapping the mind. Dog owners who adopt "beach dogs" and other canines who have missed out on these socialization periods often have to seek out professional behavioral assistance. Owners Henry and Jo contacted JCH because of the development of aggressive behavior in their dog, "Gretchen." Gretchen was described as a dog that Henry and Jo had met while on a vacation to the Costa Rican beaches. Gretchen, like Gus, had lived on the periphery of human society, as close to "wild" as a domesticated dog can be.

These were dogs that were basically "wild," dependent on humans for food, but otherwise not cared for by anyone, running free and often chased and abused by local beach bar and restaurant owners for getting into garbage cans and food stores. Henry and Jo fell in love with this younger dog, and worked hard to adopt Gretchen and get her back to the U.S., through a months-long quarantine. She was estimated by their veterinarian to be about one and-a-half years old and wasn't spayed until that vet visit in the U.S.

So not only had she experienced little and poor-quality human (and probably canine, as well) socialization during early sensitive periods in the development of her 'social brain,' but she then experienced the less than desirable environment of long-term quarantine. And all of this was during the crucial brain-development period of the first two years of life.

As is almost universal in these cases, Gretchen was diagnosed with a form of anxiety: she basically didn't know how to interact properly, with humans or with dogs, and therefore, had developed increasingly severe social anxiety. Treatment involved slowing down her introductions to family, friends, and strangers; lowering expectations for strong social bonding, at least in the near future; and implementing a positive–reward program for appropriate behavior in limited, structured, and positive social interactions. The prognosis in these cases is not good: miraculous turn-arounds have occurred with owners able and willing to do a LOT of work with their pets, but in most cases, even with hard-working owners, results are only moderate in these cases: these dogs rarely ever become open and trusting of people or other dogs.

YOU TALKIN' TO ME?

Humans communicate with their dogs, both verbally and non-verbally, from the moment that they meet. Throughout our dogs' development, including sensitive and critical periods, we socialize with them. It's no secret, then, that those who share their lives with dogs talk to them often and in ways that are typically reserved exclusively for pets and young children. Perhaps as long as people have been speaking to their dogs, they have been teased and criticized for their use of "motherese" or infant-directed speech, which includes diminutive word forms such as "walkie" and "doggie," a higher pitched tone of voice, exaggerated speech melody, repetition, and simplified grammar. But it looks like those who are saying "doggie" in dulcet tones might be getting the last laugh here – because recent scientific findings support the long-held notion that dogs, like babies, respond to this type of speech.

Visit any dog park or dog-friendly establishment and you'll be met by a symphony of singsong "Who's a good boys?" and "You wanna have a cookie?" But this silly, simplified speech is met by adoring[80] gazes: dogs who are engaged with their people and responding to this dog-directed speech. Recent brain scans of humans and dogs reveal that this connection isn't just occurring with what we can see – it's happening on a chemical level, as well. Dogs are hard wired to respond to human communicative signals and can discern what we're feeling by cueing in to what we're saying and how we're saying it. While no other dog currently approaches Chaser's record, many dogs have the capacity to understand more than 200 spoken words.[81] Human sounds and scents are particularly salient to our canine companions and recent research has shown that dog and human brains actually process the vocalizations and emotions of others more similarly than we previously thought.

Scientists at Emory University studying animal cognition used neuroimaging to assess how dogs' brains process smells. Dogs were trained to lie in an MRI machine and fMRI (functional magnetic resonance imaging) scans measured their neural responses to the smell of people and dogs, both familiar and unfamiliar. Of all the odors that they were exposed to, the aroma of their owners sparked activation of the caudate nucleus, which is commonly referred to as the "reward center" of the brain. Dogs prioritized this scent over all others.

A second study performed by researchers at Eotvos Lorand University in Budapest examined how dogs responded to certain auditory input, including canine and human barks, voices, sighs, and grunts. The study revealed that dogs and humans process emotionally laden sounds in a similar fashion. Specifically, "happy sounds" lit up the auditory cortex of both canines and humans. Despite millions of years of evolutionary separation, after approximately 30,000 years of coevolution, the unique partnership between dogs and humans has changed them both forever. These studies reveal that dogs are wired to respond to human smells and sounds – and that we still have far more to learn about the complex multimodal communication between humans and canines.

---

[80] in our anthropomorphized opinions.
[81] http://www.chaserthebordercollie.com.

DON'T MIND THE MAN BEHIND THE CURTAIN

Understanding basic behavioral science makes all the difference between effective and ineffective interactions with our dogs, including training. Positive reinforcement and punishment are both popular training paths, but they use two different pathways in the brain. Using both methods is one of the worst things that you can do to your dog; you're better off using exclusively one or the other.

Some of the most confused dogs, showing the strangest behaviors or responses to their environment, have turned out to be those that were facing significant levels of training, but receiving two different forms of training: punishment or aversive-based training as well as positive-reinforcement training. These cases puzzled me (JCH) for a long time, until the science began to emerge about differing, and often conflicting, pathways for the learning involved with each method.

Numerous studies now suggest that reward and punishment produce differential effects on procedural learning, thus, using these disparate approaches together produces less than ideal results. A 2009 study by Wachter et al. suggested that punishment and reward "engage separate motivational systems with distinct behavioral effects and neural substrates."[82]

An example from JCH's case files is Charlie, a four-year-old male Spaniel mix, who was owned by a husband and wife, John and Katie. He was exhibiting aggressive behavior towards other dogs, but highly inconsistent behavior, and was recently beginning to exhibit aggression (when he was with John) or anxiety (when he was with Katie) towards children in the neighborhood. This is when they reached out for help.

Charlie seemed like a healthy, fairly well-adjusted dog when JCH arrived at their house, but as they proceeded to go for a walk, he observed the most fascinating behavior. Charlie did not want to walk on the side on which John, or he, were walking; instead, he was always veering to Katie's side. He had John take the leash, but Charlie always preferred to veer towards Katie: He moved well behind, and the same thing happened. He finally called Katie back and had John walk ahead with Charlie, and Charlie's body language changed to appear very cowed, submissive, and quiet with John. His tail was lowered, the movements were subdued, his body was lower to the ground, and he leaned away from John. JCH pointed this out from afar to Katie, and she whispered that she didn't think that Charlie "liked" John. That was certainly interesting.

A thorough interview with Katie and John revealed some interesting details. It was discovered that John had always had dogs, big ones like German shepherds, and that he was brought up to use punishment methods: prong collars, shock collars, sharp words, swats on the nose, and so on. This was Katie's first dog, and she had diligently taken Charlie to frequent training classes, all of them using positive reinforcement. John had decidedly *not* bought into the positive-reinforcement approach and was continuing to use aversive methods with Charlie, while Katie was using treats and positive reinforcement, *often* on the same walk.

---

[82] http://www.jneurosci.org/content/jneuro/29/2/436.full.pdf.

JCH began to notice this effect more often, and he realized that many traditionally-aversive trainers were experimenting with softening their reputations by developing a new approach: reward-for-good-behavior, punish-for-bad. And this was producing some seriously confused, and quite aggressive, dogs. It has only been more recently that the neuroscience has begun to explain some aspects of why using mixed methods incorporating both rewards and punishments might be so confusing for our dogs.

JCH explains: "For Charlie, I had to prescribe the use of just one training approach. With an open mind, I suggested either one but only one, and made a pitch for positive reinforcement. They agreed to let Katie take over the dog-handling, and in my mind, the prognosis was quite good, but at last report, there was still a lot of punishment being used by John, and there had been little progress in decreasing the nervous, confused anxiety aggression on the part of Charlie."

Unfortunately, while Charlie's experience was a stressful one for him, it isn't unique – for dogs or for other animals who share our lives. Quite often, animal behaviorists and trainers are helping the nonhuman animals just as much as they're helping the humans. This applies not only to dogs and dog training, but to other species, as well. "Instead of helping people with horse problems, I help horses with people problems," explained horse trainer Buck Brannaman, whose insightful approach to horse training and equine behavior has earned him the moniker of "the horse whisperer." Dogs – or horses – with behavioral "issues" are often receiving conflicted information from the humans in their lives, either from one person who uses both reward and punishment, or from one who uses only positive reinforcement, while the other uses punishment. An animal who is taught with two different approaches and two different sets of expectations is much like a human child whose parents' opinions on discipline and boundaries conflict. Children whose parents use different approaches become insecure and uncertain of what to expect and what is expected, and often act out, just as dogs do, given the same disparities. The new dog in the forest that Alec and Ziva met was first punished by her owner (by being hit on the shoulder and shushed) and then rewarded with a treat. While the dog took the treat from his owner, he was tentative and continued to cower – he didn't know whether he was going to next be rewarded or punished, and likely didn't know why either of these were occurring, either.

The map of the canine mind has only begun to be charted, and while we know it's not flat, we have yet to understand how each region responds similarly to or differently than their wolf cousins or from humans. Understanding an animal's brain structure and their trajectory of development can help us set reasonable goals when we're helping them learn new things and throughout our lives together. The healthiest human-canine relationships take into account age-, breed-, and species-appropriate approaches. Are you asking too much of your dog too soon? Are you setting reasonable goals, given your dog's past socialization, training, and other experiences? Are you using one training method consistently and exclusively? If you alternately use positive reinforcement and punishment with your dog, you're definitely teaching them every time they're punished, but it's not the lesson that you want them to learn. You're only teaching them that you're mercurial, inconsistent, unpredictable, and unkind. If you use a treat bag for training your dog, but then punish them as well, your dog will act much like Charlie, exhibiting signs of constant insecurity and anxiety.

CHAPTER 4

What a dog knows: Analyzing sensory perception to interpret behavior

THE SIXTH SENSE

Traditionally, we've been told that there are five senses, but some neurologists now believe that there are as many as 21 different senses, including introception (balance), proprioception (spatial orientation), hunger, and thirst. The five most well-known "basic" senses – sight, sound, smell, taste, and touch – these are all readily familiar to us, particularly when we're considering another species' umwelt. But how can we know if we're sensing what another individual is sensing? And is there really a "sixth" sense that should be grouped with the most familiar five senses – a sense that we use to navigate the world and our relationships with one another? Research has shown that birds, insects, and even dogs have an ability to sense Earth's magnetic field, with one study showing that dogs prefer to defecate along a north-south axis.[83] Some scientists claim that humans have a magnetic sixth sense,[84] as well. But what is it about dogs that gives them the appearance of having the ability to sense our emotions, perceptions, and intentions? Are our dogs psychic, or is it something else? Perhaps we'll have to start with the five known senses to fully consider a sixth one.

Do a web search on "the dress" and you'll be met with an array of images of seemingly different lace-trimmed dresses: a black and blue dress, a white and gold dress, and many variations in between. While "the dress" was actually made of blue and black fabric, a polarizing debate about its colors sparked up social media during the winter of 2015. So why would millions of people disagree about something as seemingly mundane and inarguable as the color of a dress? Because 50% of respondents were certain that the dress was blue with black, while 50% of respondents were certain that it was white with gold. So how could 50% of the population be "wrong" about this? The answer lies in biology and evolution – and in the way that the human eye sees colors. "The dress" wasn't just a viral debate; it's also an excellent vehicle to explain how sensory perception can be used to interpret our behavior.

---

[83] http://frontiersinzoology.biomedcentral.com/articles/10.1186/1742-9994-10-80.
[84] https://www.sciencealert.com/scientist-claims-he-s-discovered-a-magnetic-sixth-sense-in-humans.

Color vision, which is crucial for all primates, allows for the discrimination of light that enters the eye through the lens and hits the retina, which is located at the back of the eye and contains photoreceptors (rods and cones). Rods detect shades and brightness of gray and are responsible for night vision, while cones detect color vision and are responsible for day vision. Different wavelengths are interpreted as different colors, but we only perceive the *reflected* colors. The surface of the object reflects the wavelengths back to us and our brains determine what our eyes see. A black object absorbs wavelengths, while a white object reflects them back. So what's going on with "the dress?" According to neuroscientist Dr. Bevil Conway, "Your visual system is looking at this thing, and you're trying to discount the chromatic bias of the daylight axis."[85]

According to Dr. Conway, human color vision takes into consideration the amount of daylight that's available. While human eyes, which evolved to see during the daylight hours, do well in bright sunlight, our vision is less acute during dawn and twilight.[86] Dawn has a pinkish hue, mid-day is bluish white, and twilight is reddish. "People either discount the blue side, in which case they end up seeing white and gold, or discount the gold side, in which case they end up with blue and black."

A similarly divisive sensory debate pertains to cilantro, or *Coriandrum sativum*, (which is referred to as coriander in Great Britain). Julia Child famously hated cilantro, while many chefs love this herb – and use it liberally in their dishes. But for 4-14% of the population, cilantro tastes like soap or something similarly unpleasant. Those in the "soapy cilantro" camp can't understand the perspective of those in the "savory cilantro" camp. And there's a genetic reason behind this: Most people who think that cilantro tastes "soapy" have a group of olfactory-receptor genes called OR6A2 that help detect aldehyde chemicals,[87] which contributes to cilantro's flavor. Aldehyde chemicals are also present in soap, though; so depending upon one's genetic makeup, cilantro can taste like a savory herb, like soap, or even like stinkbugs, detergent, or cinnamon, which all contain aldehyde chemicals. (It's unclear who in this study was taste testing stinkbugs, or how "stinkbug" would be a recognizable taste.) According to research pertaining to cilantro preferences, two polymorphisms (genetic variations) account for these differential gustatory differences, one in the TAS2R1 taste receptor gene, and another in the OR4N5 olfactory receptor gene.[88] In layman's terms, our evolutionary histories and recent genetic polymorphisms are strong determinants of how we will perceive the world. If you can't even see a blue and black dress as white and gold or vice versa or understand how a favorite herb tastes soapy, how can you begin to imagine the scale of sensory differences between species? With such disparities within *Homo sapiens*, imagine the world of difference between the sensory perception of *Homo sapiens* and *Canis*

---

[85] https://www.wired.com/2015/02/science-one-agrees-color-dress.
[86] http://www.bbc.com/earth/story/20150219-the-worlds-most-sensitive-eyes.
[87] https://www.nature.com/news/soapy-taste-of-coriander-linked-to-genetic-variants-1.11398#/b1.
[88] Mauer, L. K. (2011). *Genetic determinants of cilantro preference* [M.Sc. thesis]. University of Toronto. https://tspace.library.utoronto.ca/bitstream/1807/31335/1/Mauer_Lilli_K_201108_MSc_Thesis.pdf.

familiaris. If dogs were having either of these arguments, they likely wouldn't be arguing about whether the dress was blue or white or if the cilantro tasted like a savory herb or a soapy culinary component. They would likely be arguing about the smells pertaining to who had worn the dress, and where; and how the herb smelled, and where it came from.

For years, cats and dogs were succumbing to the ingestion of an unknown poison up to 48 hours after they'd consumed it. Cats and dogs who had access to areas where cars were parked would consume the antifreeze (ethylene glycol) that had leaked out of cars. Ethylene glycol, it turns out, has a sweet taste that's palatable to humans, felines, and canines – but it's highly toxic, too. Poison control centers nationwide handle more than 5,000 human ethylene glycol poisonings annually and as many as 90,000 pets are poisoned by this substance each year, as well. But once the manufacturers learned that their tasty product was leading to the loss of lives, they changed the formula, adding bittering agents to make it unpalatable for pets and people alike. Without realizing that the poison was so tasty for them, though, these deaths by poisoning may have continued unabated.

Understanding an individual's unique sensory perception is key to interpreting their behavior. Not all beings process sight, sound, smell, taste, and touch in the same way. One's umwelt, or how one experiences the world, determines how one will interact with the environment and with other beings within that environment. The bee who views the world through an ultra-violet lens will behave differently than the blind tick who senses the world through temperature and butyric acid, and the canine who experiences the world with an olfactory emphasis will behave differently from the human who relies heavily upon visual cues. Biologist Jacob von Uexküll, who focused on the intricate interplay between behavior, anatomy, and sensory perceptions, conceptualized that one's umwelt largely determines one's behavior. While humans and canines are mammals who have similar sensory systems, we process information differently, according to our own umwelts, which have been shaped by selective pressures over the ages.

Scientific studies and modern technology can help simulate a dog's sensory world, but they can't recreate it – we can't truly walk a mile in a canine's shoes, nor can they in ours. We can understand how and why a dog's senses differ from our own, but how can we truly understand how they process sight, sound, smell, taste, and touch? How can we imagine olfactory amplification of such magnitude, and that's so information-rich? How can we try to imagine pitches that are beyond our auditory range, tactile input from sensitive whiskers and paw pads, or a sense of taste that exceeds the detection of the keenest human epicurean? For canines, the hint of a scent gives as much information – or more – as a detailed handwritten note. "I was here at such and such a time; I'm healthy or sick; male or female; familiar or unfamiliar…" A dog's sense of smell can be more accurate than intelligence gleaned from espionage or personal information acquired from data mining. All of that information, in one smell, however brief, like data stored on a tiny computer chip – how can we comprehend it, and why is sensory perception so important for interpreting behavior?

The anatomy of an organism's eyes, nose, ears, and mouth; what is salient to an organism; and how that information is processed are all important factors in how an individual will behave. Anatomy's role in sensory perception is more complex than a cursory glance might imply: anatomical placement, size, and range of movement convey much about an animal's perceptive skills. Are the organism's eyes forward-facing for stereoscopic or binocular vision, as in many predator species, including the felines and canines? Are the organism's eyes placed on the side of their head, allowing for greater peripheral vision, as in many prey species, including equines and bovines? Is their nose prominently placed on their face? Can their ears rotate independently? Do they have whiskers that are rich in sensory cells?

What is salient – or of most importance – to an organism also determines how an organism will behave in its environment. Ethology founding father Niko Tinbergen pioneered the term "supernormal stimuli," which refers to an exaggerated version of a salient stimulus. This exaggerated version of the stimulus receives an exaggerated behavioral response, even to the eventual detriment of the individual – and disregarding the evolutionary purpose of the natural stimulus. For example, a songbird might choose a larger, more loudly colored, artificial egg to incubate over their own, because it has the characteristics of one of their own eggs, but to a greater degree, thus making it more stimulating than their own egg. And a greylag goose might disregard its own eggs in the presence of a volleyball, which was "more stimulating" than its own eggs. Or a beetle might attempt to copulate with a discarded beer bottle because it was more jewel-like than the carapace of an actual beetle.[89] In the 1980s, researchers Darryl Gwynne and David Rentz discovered that male Australian jewel beetles (*Julodimorpha bakervelli*) were attempting to copulate with beer bottles that had similar coloration and tubercle placement as female carapaces. The males even continued to "copulate" with the bottles even as they were attacked by ants. Neither of these behaviors would actually benefit the reproductive success of the bird, whose offspring would die; or the beetle, whose reproductive efforts would yield no offspring. Egg size and pattern are salient to many bird species; carapace size and coloration are salient to many beetles; and for dogs, humans – including their sounds, smells, and facial expressions – are particularly salient.

When it comes to perception, though, canines don't have the keenest senses, comparatively speaking; felines outrank canines in overall sensitivity. Felines have a sharper sense of smell, more acute night vision, and a much keener sense of hearing, as well, with ears that can rotate 180 degrees. But when it comes to humans, canines have a distinct evolutionary advantage over felines – an intuition that's so keen that it could almost be considered a "sixth sense." Dogs have co-evolved with humans for 32,000 years, and perhaps longer, while cats have likely cohabitated with humans for only 9,000. As a result, dogs are considered to be domesticated, while cats are only considered to be "semi-domesticated." And it was during this long domestication process that dogs developed a sense of human intuition – a sense that cats may further develop, as well, given sufficient time.

---

[89] Gwynne, D. T., & Rentz, D. C. F. (1983). *Beetles on the bottle: Male buprestids mistake stubbies for females (Coleoptera)*. Canberra, Australia: Department of Zoology, University of Western Australia.

SEEING THE FOREST FOR THE TREES

Humans, like all primates, are visually oriented beings – our sense of sight is an important part of our evolutionary and ecological heritage. Among eutherian[90] mammals, primates have unique color vision called "trichromacy," which means that we possess three types of color receptors (cones). These three cones, which are green-, yellow-, and violet-sensitive, transmit color information and enable us to perceive the three primary colors[91] and upwards of 10 million different colors. Most birds, reptiles, and fish have tetrachromatic vision, meaning they have four types of cones in the eye to transmit color information, while all mammals excepting marsupials and some primate species, have dichromatic vision. Dichromats, including canines, have two cone cells in their eyes and thus they see color differently than trichromats; they're said to be "color blind" in the same way that some humans have color blindness. But this doesn't mean that dichromats have a loss of experience; having known no other visual reality, they simply interpret visual data differently than trichromats do, and both assume that their umwelt is reality.

Dogs experience difficulty distinguishing between green and red, seeing both as variations of yellow, because they only have two types of cone cells. Their vision is similar to how "color blind" primates would process color information. So why do primates excel in this type of vision, in comparison to other eutherian mammals? Trichromatic color vision has evolved convergently among distant primate species and scientists have long thought that primates' ecological niches provided strong selective pressure for trichromatic color vision. Primates' color vision was thought to have evolved from searching for food among the forest and recent research[92] buoys this hypothesis.[93] Possessing the ability to discern reds from greens is advantageous for spotting nutritionally important foods, such as young leaves and ripe fruit, from a distance. It is thought that primates from many generations ago, much like their extant ancestors, lived in forested environments where those who could discern these red-green differences had a selective advantage over those who could not.

Over millions of years, those primates that could discern reds from greens had more reproductive success, while those that could not discern these colors had comparatively fewer offspring, eventually disappearing from the primate family. Dogs, however, did not experience this selective pressure; as predator-hunters, rather than omnivorous hunter-gatherers, discerning movement was more important than discerning red-green. Canines can discern the movement of an animal before humans can, just as Ziva spotted the moving wolves before Alec did; but dogs would have no evolutionary reason to detect a red apple from within green foliage. Dogs' vision

---

[90] Eutheria is one of two mammalian clades with living members that diverged in the Early Cretaceous or Late Jurassic periods.

[91] The free dictionary.

[92] Dominy, N. J., & Lucas, P. W. (2001). Ecological importance of trichromatic vision to primates. *Nature*, *410*(6826), 363–366.

[93] Surridge, A. K., Osorio, D., & Mundy, N. I. (2003). Evolution and selection of trichromatic vision in primates. *Trends in Ecology and Evolution*, *18*(4).

is superior to humans at dusk and dawn and with moving objects, and their low-resolution vision is better in comparison to humans' vision, but for color detection, it is the primate family that can truly see the forest for the trees. It should come as no surprise, then, that humans love optical illusions and images that contain more than one possible reality. Visual trickery manipulates our perception – what we're seeing disagrees with our "reality." But what's novel or hard to decipher can also be salient, harkening back to a time when our ancestors had to find ripe, nutrient-rich fruit or young new leaves in a sea of inedible foliage. While primates are very visually oriented, however, a dog's world is primarily comprised of smells and sounds – and with a range far beyond human capacity. Our sensory differences are just as important as our sensory similarities – they not only shape how we view the world, but how we process information and interact with other organisms.

Dogs, for example, can be spooked when coming through a blind corridor, just as Ziva was as she came through the foliage as she walked with Alec. Recent legal cases have addressed the detrimental effects of blind corridors at dog parks.

The concept of the umwelt is imperative to understanding how the dog views its world very differently than we do. It's not just from the perspective of a different sensory system bias (for us, we rely primarily on vision, while dogs rely primarily on their sense of smell), but from their unique point of view, as well. There have been several legal cases of dog/dog and dog/human bites that revolve around the fact that many parks and walking areas, while seeming open to the humans towering above, are actually very closed and tunneled and confusing or stressful places for the dogs.

Designing a dog park without taking into consideration the dog's point of view (several feet below the human's eyes, ears, and nose) is highly problematic. Often high grass is not mown, which produces an open parkland effect for humans, but creates closed-off tunnels, blocking vision and channeling an overload of smells, for the dog. There's often narrow trails maintained through high grass, or narrow bridges over small creeks, where dogs (many of them with dog-dog or dog-human sensitivities already) have to come within close proximity to their fear-triggers. This increases anxiety in the dog. They wonder, "Who's going to come around the next corner and how are they going to act?". The owner who is unaware fails to understand the reaction of their pets. In these legal cases, giving this information to a hearing examiner or a judge often provides a valuable perspective for them to at least understand *why* the dog has behaved the way it has…and the question of fault is theirs to decide.

What a dog can – or can't – see strongly impacts their behavior. While there's still relatively little literature about the subject, dogs have also been known to attack homeless people, both because of the visual cues (including a novel rocking motion from those with mental illness and the fact that the person is sitting at the unusual height of their line of vision) and because of their odor (unbathed, and often smelling of alcohol.)

In Seattle, where JCH lived and practiced as an animal behaviorist for 27 years, he has seen two cases in which dogs walked by their owners, relatively stable, well-behaved pets, have "inexplicably" turned on the street and attacked a person sitting quietly. By "attacked," he describes how the dog turned without warning and bit them hard enough to punch a hole in their clothing or lightly break the skin.

When reported to me this way, the cases were mystifying, but again, the circumstances are everything. In both cases, these were relatively highly reactive dogs who had experienced recent periods of stress (one was a recent adoption from a shelter, and the other had experienced the recent departure of a long-term human caretaker [the boyfriend of the owner]). They indeed had no history of aggression, especially towards humans, so from that point of view, the bites were surprising. But in both cases, the bites were directed towards homeless men, laying on a sidewalk (which is not a normal location or position for humans, at least from a dog's perspective), behaving 'erratically,' according to witness accounts, and with a strong smell, again according to witness accounts. From my perspective, I cannot blame a mildly wound-up dog, passing closely to a human in an unusual posture, acting and smelling unusual, and who, I suspect, reached out to the dog, or otherwise made an unexpected movement, for turning and biting lightly (the bites as described were unquestionably bites, but were of only light to moderate intensity). Again, it's my hope that my ability to provide a perspective from the dog's world, rather than the human' world, may help explain, if not excuse, these undesirable interactions and perhaps provide an opportunity for dogs trapped in these situations to be given a second chance. But for sure, these situations occur in the dogs' umwelt and not in ours, and hence can be mystifying for owners.

THE NOSE KNOWS

As primates, humans are visually oriented, but canines experience the world primarily with their noses. Just as primates faced strong selective pressure for trichromatic vision, canines faced strong selective pressure to be able to detect prey – and other members of their social groups – over long distances. The evidence for this selective pressure is mapped in our brains: The human brain is roughly ten times larger than the canine brain, but the portion of the canine brain that's dedicated to olfaction is 40 times greater than it is among their human companions. Humans have six million olfactory receptors in their noses, but dogs have 300 million olfactory receptors. And while dogs have hundreds of millions of receptors, the domesticated cat has a sharper sense of smell still. But when it comes to human-related scents, dogs excel once again: it's in that period of domestication that dogs maintain the upper hand on cats.

Dogs have a more acute sense of smell than humans do, but do they process, categorize, and remember smells similar to humans? Olfactory memory is one of the most primitive and powerful kinds of memory, and a recent study found that dogs' sense of smell is even more complex than we realized. There is now evidence that dogs can categorize unfamiliar smells based upon their prior experiences. The study, published in *Scientific Reports*,[94] provides evidence that dogs categorize smells that they hadn't previously encountered using a system similar to humans: they rely upon

---

[94] Wright, H. F., Wilkinson, A., Croxton, R. S., Graham, R. C., Hodkinson, H. L., Keep, B., et al. (2017). Animals can assign novel odours to a known category. *Scientific Reports, 7*(9019).

their prior olfactory encounters. Dogs not only placed scents in "similar" categories, but they remembered them six weeks later, too, revealing that olfactory memory is particularly important to dogs as well as to people. For humans, olfactory memories are referred to as the "Proust Phenomenon" after French writer Marcel Proust, whose pen was at times equal parts poetic and neurological. Memories of an event can be more vivid when there is an odor linked to it. Perhaps this is why our dogs linger so long in familiar areas during their walks – they're not only reading the information, but they're also having vivid memories of the last time that they visited the area.

When Jack is walking along one of his favorite routes, he will often pause and then propel his entire face into a thicket or a large tuft of grass, inhaling some amount of information that we can't sense. After many moments, he'll continue to sniff deeply, even if his leash is pulled on strongly. He's likely smelling which of his friends had also been there, and how they're doing; which unfamiliar dogs visited, and perhaps if any birds, bunnies, or squirrels had recently passed through that patch of earth. For Jack, like all dogs, a walk is a sensory experience to be savored; the smells and the socializing are the two best parts of the experience.

Dogs are at the heart of a $60 billion a year industry that includes dog parks, doggie daycares, animal shelters, and animal rescues. But what do any of these places smell like to a dog? What does anything smell like to a dog? While dogs are the epicenter of the booming pet industry, how can we truly market products for them if we can't fully understand their sensory worlds? Dogs are so much better at smelling than humans for two reasons: first, the olfactory portion of the canine brain is much better developed than the corresponding portion in the human brain; and second, they have a greater density and number of sensory cells. Humans smell in parts per million, while dogs smell in parts per trillion. Why is that such an important discrepancy? In her book *Inside of a Dog*, dog cognition researcher Alexandra Horowitz explains that a dog could detect one teaspoon of sugar in one million gallons of water. Just how much water is that? One Olympic regulation sized pool holds 500,000 gallons of water; a dog's sense of smell could detect only one teaspoon of sugar in something twice that size. This is great news for humans like Luke Nuttall, who has Type 1 Diabetes. His dog, Jedi, alerted Luke that his blood sugar had dropped, even when the glucose monitor failed to do so.[95]

So imagine what an actual swimming pool, with all of the commingled scents of humans, would smell like for a dog; imagine what a person would smell like, when they came home from work or an extended absence, smelling of everywhere they'd been, who they'd been with, and how healthy they were at that time. While it's hard to understand what it would be like to have a sense of smell that's exponentially better than our own, it's clear that dogs are smelling far more than humans are – and pet owners and scientists alike have found great value in this. Human scientists detect sickness, including cancer, using laboratory results. So what does a sick person smell like for a dog, in comparison to a healthy one? Illness alters the body chemistry, so while it's impossible to ascertain precisely "how" a sick person smells to a dog, we

---

[95] http://news.nationalgeographic.com/2016/03/160319-dogs-diabetes-health-cancer-animals-science.

can assume that a sick person's odor is qualitatively different from that of a healthy person. Dogs can sense volatile organic compounds (or VOCs), which are released by cancerous cells, aiding in the detection of an array of human illnesses. Among the diseases that dogs can sense are cancer, diabetes and low blood sugar, narcolepsy, epilepsy and other seizure disorders, and migraines. They can even smell pain and fear. Just how accurate are dogs with detecting these differences? Dogs like Lucy,[96] a Labrador retriever/Irish water spaniel mix, have shown that not only do dogs have the ability to smell cancer cells, but when properly trained, they can detect these deleterious cells with an astounding 95% success rate.

But do all dogs have such a keen sense of smell, or is there a relationship between form and function? Long before the ancestors of modern man began to selectively breed the ancestors of domesticated dogs, canines had a long snout and upright ears. Over years of artificial selection, many dog breeds are characterized by protracted faces with short snouts, long, floppy ears, or both. But there's a reason why ancestral dogs – and dogs who are successful hunters – have longer snouts, comparatively speaking. Understanding the dog form and function of a dog's snout can help us better understand their behavior. Dogs like bloodhounds, who have long snouts, have a better sense of smell because their noses contain more olfactory glands than a short-snouted dog like a Pekingese. And while there's certainly individual variation, a "scent hound" like a bloodhound does have certain morphological advantages, such as a longer nose, scent-trapping skin folds, and long ears, which also collect scent. A dog's sinuses are designed to capture air and sensation; not only are the tremendously bigger than our own, but their comparative brain real estate that's devoted to olfaction is also much larger than a human's olfactory brain center. From nose length to neurology, their detection system is far more elaborate than ours. Contrastingly, a dog's vision isn't as sophisticated as a human's, nor is the visual component of their brain as well developed as the human brain.

HEART OF HEARING

Dogs, who have exceptional hearing in comparison to humans, are actually born deaf and only begin to hear around three weeks of age. In comparison, newborn human babies are born with the ability to hear and will exhibit a startle reflex at sudden noises. Over the course of their development, however, dogs' sense of hearing exceeds that of their human companions, and on a hierarchical scale of sensitivity, their sense of hearing is second only to their sense of smell. On the higher end of the frequency scale, humans typically can't hear anything above 23,000 Hertz, while dogs can hear as high as 45,000 Hertz. Cats, in comparison, can hear 100,000 Hertz. On the lower end of the scale, humans can hear noises as low as 64 Hertz, but dogs can only hear 67 Hertz. Dogs can hear termites in the walls, fluorescent lights, and the slightest movement of a cheese wrapper or treat package. While humans only have six muscles that control ear movement, dogs have 18, allowing for a wider

---

[96] https://www.cnn.com/2015/11/20/health/cancer-smelling-dogs.

range of movement and more efficient reception of sound waves. Dogs' ears can move independently of one another, allowing them to attend to and filter out different sounds simultaneously. Cats have 30 muscles that control ear movement, allowing for even greater movement and capturing even more sound in their funnel-shaped ears. A dog can rotate one ear in one direction, even while their eyes and other ear are otherwise engaged. But it's with humans where dogs appear to find the most engagement.

Dogs appear to be able to tune in specifically to the human voice – this is a sound that's particularly salient to them. And studies provide evidence that dogs process speech in a way that's similar to how humans do,[97] processing words and tone of voice in separate hemispheres of the brain. Researchers used functional magnetic resonance imaging (fMRI) to measure how dogs neurologically categorized intonational and lexical information. When the dogs in the study registered tone, the fMRI revealed that their right hemispheres showed a response, and when they registered known words of praise, their left hemispheres responded. The left hemisphere is where humans process language, as well. The results showed that dogs could decipher not only individually known words, but intonation, as well, revealing that what we say and how we say it can both be important when communicating with canines.

Dogs and humans have similarities in their neural architecture. With our co-evolutionary histories, it's not surprising that human voices are particularly salient to dogs, nor that they process human languages and sounds in a manner that's neurologically similar to humans. Dogs are uniquely tuned in to humans on all sensory levels, responding to our emotional states, health, and gestures, and they're particularly tuned in to the sound of human voices. But what's happening neurologically? Humans process language – including recognizable spoken words – in the left hemisphere of their brain, while the right hemisphere processes other information, including the emotional tones of voices. Humans process the content of sounds in the ear opposite of the associated hemisphere because the strongest sound pathways connect to the opposite hemisphere. Thus, the left ear would connect to the right hemisphere, and the right ear would connect to the left. When humans process recognizable auditory language, they exhibit a right ear bias (left hemispheric bias) and when they process the emotional information in voices, they exhibit a left ear bias (right hemispheric bias). Research now shows that depending upon the content of the speech, dogs also show these orienting asymmetries, revealing that dogs' hemispheric language responses are similar to humans' responses.[98]

GOOD VIBRATIONS

Born deaf and blind, the first sense to develop in a newborn dog is its sense of touch. From whiskers to toes, a dog's body is covered with nerve endings that are particularly sensitive to touch. Unlike humans, cats, dogs, and multiple other species of animals,

---

[97] http://science.sciencemag.org/content/353/6303/1030.
[98] Ratcliffe, V. F. (2014). Orienting asymmetries in dogs' responses to different communicatory components of human speech. *Current Biology, 24*(24), 2908–2912.

including bears, rats, and seals, have coarse, deep-rooted whiskers on their faces that provide important information about their environment. Whiskers, or *vibrissae*, as they're scientifically called, provide important sensory input for cats and dogs, as their follicles are dense with touch-sensitive neurons. Whiskers enable improved nocturnal navigation, as they can detect changes in air currents. With cats and some smaller dog breeds, these tactile hairs help determine whether they can fit into certain spaces by vibrating when they brush against an object. If their whiskers can't fit into a space, then their bodies likely can't, either. *Vibrissae* derives from *vibrio*, which is Latin for "to vibrate." While humans rely on their sensitive fingertips for much of the information they receive about touch, whisker tips encode important information for cats and dogs. Vibrissae function much the same way that a blind person uses their cane, sweeping the area to make a mental topographical map, but with more accuracy, as vibrissae are part of the animal and not merely a tool. When an animal is approaching an object or a new area, they'll "sweep" with their heads, and their vibrissae will vibrate along the surface, providing feedback about an object's topography. Is the object or area smooth or rough? Will they fit through the entryway, or is it too small? Olfactorily-oriented dogs rely less on their eyes than they do on their sense of smell and touch when they're exploring objects that are up close, and vibrissae provide these answers for them (Fig. 24).

How important are our dogs' vibrissae? Almost 40% of the portion of a dog's brain that registers tactile information corresponds to the face, and the majority of that corresponds specifically to the regions of the face where vibrissae are located. That's a lot of neural real estate – and this strongly suggests that vibrissae are very important for whiskered animals. Dogs and cats whose whiskers have been trimmed or removed can struggle to navigate their environment, especially during lower light. Whiskers can also indicate a dog's emotional state, flaring out and pointing forward when they're feeling threatened. From navigation to object avoidance, these "feelers" provide a rich amount of sensory information for dogs. With how sensitive they are, it isn't surprising that dogs and cats can get "whisker fatigue," which often manifests as a reluctance to have their whiskers stimulated, for example, by avoiding a bowl or dish that has high sides.

IT'S A MATTER OF TASTE

Sweet. Savory. Smoky. Bitter. Buttery. Bland. Fishy. Fruity. Full-bodied. Humans have hundreds of words to describe the complex flavor profiles of our food. Socially, culturally, and evolutionarily, our sense of taste is very important to us. When we describe a fine wine, we might call it oaky, full-bodied, with hints of nut and chocolate, spicy, peppery, or sweet. We might mention that it is aromatic, and swirl the glass and inhale its scent, but this is all to open up the wine and enhance its flavor profile. A "refined palate" is a highly esteemed quality to humans, evidenced by the $800 billion a year restaurant industry[99] and the growing popularity of cooking and baking television programs. To have "good" taste is more than a compliment to one's

---

[99] https://www.restaurant.org/News-Research/Research/Facts-at-a-Glance.

FIG. 24

Canine vibrissae and feline vibrissae. *Sketches by Arianne Taylor.*

style; it's a literal classification of one's gustatory gifts. In 2015, 779 million gallons of wine were consumed in the U.S.,[100] but there are only 147 master sommeliers (those who have the most refined palates and wine knowledge) in the entire United States.[101] Taste has cultural, social, dietary, and biological connotations – and having "good" taste is elite. Taste tells us what's safe and pleasant to eat and what's not. For our ancestors (and for those of us who are still brave enough to eat what we forage in the forest), foods that were particularly bitter or strong tasting could be toxic, and ingesting them could lead to gastrointestinal distress or even death.

Surprisingly, while dogs have a superior sense of smell in comparison to humans, their sense of taste isn't as acute. While the average human has 9,000 taste buds, dogs only have 1,700 – and cats a paltry 473. Even with this knowledge, humans still view food – whether it's for our consumption or for our pets – through our umwelts, and with our hierarchy of flavor taking precedence over scent. Our dogs do have the same taste senses that humans do, although they aren't as strong. They can taste salty, sweet, bitter, sour, and the savory umami. For dogs, salt enhances the taste of sweet.

The pet food industry has invested millions into developing foods that are nutritious and delicious, but what would our pets ask for, if they could? While humans focus on the ten different intensities of the shellac that might be put on pet food, or the flavor profile of salmon with sweet potato, that's a marketing tactic that reflects our human sensory realities. Our animals would want food that was aromatic; unlike humans, taste would be a secondary consideration for them. But we're stuck in our umwelt and we don't want pungent smelling food, because that's repulsive to us. Take Limburger cheese, for example (or don't – that's what sparked the Great Cheese War of 1935.) Limburger is a semi-soft cow's milk cheese that's particularly pungent because it's a washed-rind cheese with bacteria growth on its outside. This produces added odor. This would be highly palatable to canines, but to humans? Not so much.

In the winter of 1935, a physician from Independence, Iowa prescribed pungent Limburger to a patient whose sinuses were stuffed up, believing that the cheese's strong aroma could alleviate the congestion. The prescription was as unusual as it was convoluted to get it to the patient. The doctor ordered Limburger from its headquarters in Monroe, Wisconsin, but the cheese never arrived at its intended destination 127 miles away because the postman handling the odiferous "medicine" was offended by its smell. He refused to deliver the Limburger, stating that its aroma was "objectionable" and could "fell an ox at twenty paces." This incited an all-out fromage fight that was finally resolved when people agreed that they could forgive Limburger's aroma because its earthy, savory taste was more than palatable. This never would've been an argument, if dogs were to decide. That strong-smelling cheese probably would've been consumed long before it crossed state lines.

Why the disparity between sensory priorities? It's in our genes, our evolutionary histories, and our umwelts. Our ancestors were originally herbivorous primates, and pungent smelling food could have indicated something that was rotting and unsafe to eat. Not so

---

[100] https://www.wineinstitute.org/resources/statistics/article86.
[101] https://www.mastersommeliers.org/about.

for the meat-eating ancestors of modern domesticated dogs. Many believe that it was the garbage of our camps – marked by its pungent odor – that first attracted dogs to humans. Dogs' ancestors hunted, but they likely also fed opportunistically on carcasses scavenged from others animals' kills. Thus, pungent scents indicated a high likelihood of edible rewards. We erroneously think that taste is as important to our dogs and cats as it is to us, but it's not. For our feline companions, water content, smell, and texture are what they're really seeking for dinner. Humans want food that's "flavor blasted," but that's a marketing hook for us, not an enjoyable culinary experience for our pets. That "flavor blasting" doesn't have the same impact on a cat that it would on a human. Cats, unlike humans, don't always care about the taste of their food, as long as it passes the "smell test."

Our sensory perception is key to our behavior. A meadow filled with disparate species such as a bee, a tick, a bat, a rattlesnake, a human, and a dog would be interpreted completely differently by each species. For cats and dogs, smell is more important than sight and scent is more important than taste. These differences in our evolutionary histories and genetics impact how we perceive and behave within our environments. Just how different is a visit to the dog park for a human and for a dog? It would smell like a lovely day for us, and we'd cue in to the visual aspects (trees, other people, other dogs, the lay of the landscape), and perhaps hear pieces of conversation on the wind. For a dog, however, there would be the overwhelming odor of pee mail – the calling cards of one hundred dogs that had recently moved through there. This is a perceptual comparison that's far beyond arguing whether a dress is black and blue or white and gold; it's like asking what color the dress is and the respondent replying, "I don't really care, but it smells like it was in the dog park today."

Understanding not only a dog's perspective, starting several feet below our own, but the framework of our dogs' sensory perception, as well, can help us interpret their behavior in meaningful ways – and enable us to keep them motivated during the learning process, too. We know that certain senses (such as smell and sound), are more important to dogs than their other senses are, and we know that dogs prioritize certain smells and sounds, as well. Our praise lights up our dogs' reward centers, so it's no surprise that many dogs can be successfully trained with praise as a motivator for the desired behavior. Our dogs are also particularly interested in pungent treats, and having these on hand can make the difference between success and failure, especially when a dog is first learning a new desired behavior. The secret to a successful canine-human relationship is putting yourself into the mind, and senses, of the dog, to understand what motivates them, or terrifies them, or satisfies them. Dogs appear to have a "sixth sense" when it comes to reading humans, and by viewing the world through their unique sensory perspectives, we can facilitate more effective interspecies communication.

CHAPTER 5

The emotional animal: Using the science of emotions to interpret behavior

THE AGE OF DENIALISM

In the not so distant past, the idea that animals were capable of feeling "emotions" was unorthodox, at best, and blasphemous, at worst, and ethologists have long struggled with conflicting schools of thought. While Darwin was one of the first scientists to write about the emotional lives of nonhuman animals, most early scientists and philosophers (and philosopher-scientists) were animal emotion denialists, and none of them more so than Rene Descartes. Descartes postulated that animals were "automata," beings that behaved in ways that gave the false appearance of emotion and consciousness, but who were, in fact, merely mechanistic. But in the centuries since the "dark ages" of animal science, new hypotheses have been presented, suggesting that there's not only ample scientific evidence of emotion in animals that are seen as "complex" and "charismatic," such as dolphins, nonhuman apes, elephants, and dogs, but even among insects, who have some of the most simplistic neural circuitry. So do nonhuman animals experience emotion in the same way that humans do, or are their emotional experiences qualitatively different from our own?

For centuries, scientists have debated what emotions are, and whether nonhuman animals – and dogs, specifically – could share the entire spectrum of emotions that humans do. Early on, Darwin made a strong case for the existence of emotion among nonhuman animals, using dogs as examples throughout his book, the *Expression of Emotion in Man and Animals*. And while dogs have co-evolved beside us for millennia, the expression of emotions among dogs is not entirely equivalent to the expression of emotions among humans. Dogs can express excitement, distress, contentment, fear, anger, joy, shyness, and love, but most ethologists today believe that they stop short of expressing shame, guilt, pride, and contempt. This is because there are two fundamental categories of emotion: physiological, or "base" emotions, also called primary emotions, and secondary emotions. Primary emotions are fundamental biological responses and are considered to be universally expressed, while secondary emotions such as jealousy, guilt, envy, and shame are believed to require higher cognitive capacities that are only exhibited in a few species, including humans and nonhuman primates. The question today isn't whether nonhuman animals have emotions, but whether they have the *same* emotions that humans do. So, *do* dogs have the same emotions that we humans do? Well, yes…and no. From Descartes to Darwin to the discoveries of present-day researchers, it's been a circuitous route from emotional denialism to emotional acceptance. Recent research points toward the likelihood that nonhuman animals experience emotion *similarly* to humans, but not entirely the *same*. So how did we get from denial to acceptance?

PARADIGM SHIFTS

When archaeologist and paleoanthropologist Louis Leakey called on Jane Goodall to begin her seminal ethological work in Africa in November 1960, scientists had long-held preconceived notions regarding the behavior and cognitive capacities of *Homo sapiens* vs. all other animal species. The findings of ethologists – especially fledgling ones like Goodall – went against centuries of anthropocentric[102] doctrine that man was not another animal, but a being above and removed from the animals. The prevailing paradigm of the time held that a slew of behaviors "uniquely" defined humans – behaviors that supposedly reflected our "superiority" in comparison to the animals. Among those behaviors were deception and perspective taking (often collectively referred to as "theory of mind"), culture, morality, personality, emotions, and tool use. "Man the hunter" and "man the tool user" were often synonymous, but Goodall, who didn't have a background in science, didn't allow that to influence her observations of chimpanzees in Gombe, Tanzania. Armed with a notepad, incomparable observational skills, and an open mind, the young ethologist ventured into the wild. During her early observations of chimpanzees (*Pan troglodytes*), she saw a male chimpanzee who she named David Greybeard using some kind of an apparatus to extract ants out of a mound of dirt. After David had finished his arthropod appetizers, Goodall approached the mound and saw that the chimpanzee had been using long blades of grass to extract (perhaps by agitating them into clinging to the grass) the ants. *David was using the blade of grass as a tool* (Fig. 25).

Goodall reported her findings to Leakey, who had long been her mentor – and who realized the magnitude of the discovery. "Now we must redefine 'tool,' redefine 'man,' or accept chimpanzees as humans," he famously said. While the discovery was groundbreaking, no one wanted to redefine anything – yet. The goal post was merely shifted to specify that "tool manufacture and modification" was now the standard to demarcate man from animals. But Goodall soon found that chimpanzees didn't just use tools; they made them, too. We've now found that a host of other species use and manufacture tools, as well, including orangutans, crows, ravens, elephants, gorillas, sea otters, dolphins, octopuses, and even ants. During her research and the research of subsequent ethologists, a growing body of evidence was found to support the existence of all of the so-called unique human behaviors and capacities, including the presence of emotions among nonhuman animals. While there had long been

---

[102] viewing humans as the central and/or most important element of existence, placed above the other animals.

FIG. 25

Jane Goodall made the groundbreaking discovery that chimpanzees, like humans, make and use tools. *Sketch by Arianne Taylor.*

anecdotal evidence of emotion among nonhuman animals, shifting from long-held beliefs – including the Cartesian view that animals were unfeeling, unreasoning automata – didn't come easily or quickly.

So if people were so reluctant to say that animals can experience emotions, what does it mean to have them? Emotions are an integral part of the human experience, and the inability to express or interpret them makes it difficult to communicate with others. People with certain neurological differences, such as those who have been diagnosed along the autistic spectrum or those who have sustained trauma to certain parts of their brain, experience deficits in displaying or interpreting emotions. It's important to be able to convey our feelings through expressions, as much of what we communicate with one another is conveyed with gestures, body postures, and expressions. This is evidenced by the use of emoticons with text messaging – a brief response could be interpreted as terse, but when accompanied by a "smiley face," the recipient is reassured of its context. The dog version of this would be using "friendly" visual gestures, such as a tail wag, breathy pant or laugh, or a play bow.

So what does "emotion" mean? According to Merriam-Webster, it's an "affective aspect of consciousness or a state of feeling: a conscious mental reaction, such as anger or fear, that's subjectively experienced and usually directed toward a specific object and typically accompanied by physiological and behavioral changes in the body."[103] We've likely had a word for emotion for as long as we've had language itself. The current word

---

[103] https://www.merriam-webster.com/dictionary/emotion.

to denote our expressive experiences finds its origins in the French *emouvoir*. Roughly translated, this means "to stir up," which is a good place to start when you're talking about some of our most ardent experiences. Did early man consider emotion to be solely in the domain of humans, or did he instead recognize that emotional experiences were shared across the animal kingdom? So important was this question that Charles Darwin devoted an entire book to the topic.

When it takes a dozen years or more to write each of your first three groundbreaking tomes, one might find it difficult to write a fourth book, but *The Expression of Emotions in Man and Animals* did not disappoint. Darwin's first book, *Voyage of the H.M.S. Beagle*, was published in 1845; it was 14 years before his next major work, *The Origin of Species* was published. A dozen years later, he followed up with *The Descent of Man* (1871), and then only one year afterward, published *The Expression of Emotions in Man and Animals* (1872). While all of Darwin's works were revolutionary during their time, it's the last of Darwin's four books that sews together his theme of biological continuity, and that the difference between man and nonhuman animals is one of degree and not kind. Recent research has vindicated and validated Darwin's hypotheses while rejecting those of Descartes and Charles Bell, who argued that a higher power bestowed humans with emotions. In his book, *Anatomy and Philosophy of Expression* (1824), Bell declared that humans had "divinely created" muscles to depict their feelings and that "Expression is to the passions as language is to thought," not considering that expressions, emotions, or even language capacities could be demonstrated beyond the human domain. But that's exactly the argument that Darwin presented. From eyebrow positions to the neurology of movement, *The Expression of Emotions in Man and Animals* outlined the emotional similarities among man and animals. Darwin wrote: "…the young and the old of widely different races, both with man and animals, express the same state of mind by the same movements." Drawing from multiple sources, including his own infant son, William Erasmus, and the responses from an 1867 questionnaire about emotional expressions among different ethnic populations, Darwin detailed emotional expressions, including, but not limited to, grief, despair, anxiety, dejection, guilt, pride, disgust, surprise, fear, patience, astonishment, shame, and modesty. The thread that sewed this book together was the continuity of expression between humans and nonhuman animals: human emotional expressions and animal emotional expressions were linked. Darwin showed the connection between expression, mental states, and neurology: the neurological organization of movement. Darwin came from a family of dog lovers, so it's not surprising that he used canines as examples in his book about emotional expression. But even before this tome, he'd been fascinated by dog behavior. While the birds and the beaks of the Galapagos and South America were considered to be the catalyst for his first book, dogs had been on his mind since the 1830s and were the starring species of his fourth book. When Darwin returned to England after his five-year sojourn aboard the *Beagle*, he discovered that one of his dogs (who was typically aggressive with people that he didn't know) seemed to remember him and still display affinitive behaviors. Darwin realized that there was something more to this – this dog wasn't a passive vessel, as scientists like Nicolas de Malebranche had

Paradigm shifts

proposed. This dog remembered him and used that knowledge to respond to him accordingly. Prior to Darwin, many scientists, including de Malebranche, believed that animals "ate without pleasure, cried without pain, and acted without knowing…" that they desired, feared, and knew nothing. But in the expressions of eyebrows, mouths, and eyes, and in the remembrance of individuals such as his dog who greeted him warmly even after a five-year absence, Darwin saw something else: that there was an undeniable continuity among all animals (Fig. 26).

Just as Goodall's findings with chimpanzees opened up a dialogue about the capacities about nonhuman apes, so did Darwin's *Expression of Emotions in Man and Animals* open up the door to discuss the emotional experiences of nonhuman animals. Today, we even discuss the existence of "emotions" in organisms as simple as insects. In a recent Ted Talk on neurology, chemicals, and emotion, researcher David Anderson noted that there's even evidence of emotion in fruit flies (*Drosophila*). Anderson and his colleagues built a device to "irritate" fruit flies by delivering puffs of air to them. After several consecutive puffs of air, the flies became persistently hyperactive, with the frequency and intensity of the puffs increasing the duration of their hyperactivity. The flies appeared to be acting "frustrated," just as humans might – and there's a chemical reaction similarity between flies and humans, too. According to Anderson, "Flies have dopamine and it acts on their brains and on their synapses through the same dopamine receptor molecules that you and I have. Dopamine plays a number of important functions in the brain, including in attention, arousal, reward, and disorders of the dopamine system have been linked to a number of mental disorders, including drug abuse, Parkinson's disease, and ADHD."

A growing number of scientists no longer refer to the animals they study as unthinking, unfeeling beings, or merely as numbers, rather than as individuals with distinct

FIG. 26

One of Darwin's dog sketches; Darwin saw undeniable continuity among all animals.

personalities. While many scientists are very careful to avoid anthropomorphism, dogs, like humans, and like all animals, have dopamine and all experience similar chemical experiences. Long before she knew about this chemical continuity, Goodall saw behavioral familiarity with the primates that she worked with. She named the chimpanzees that she observed, which was against the doctrine of the time, but helped bring these individuals to a larger audience and helped change the tone of subsequent research methods. Telling stories about David Graybeard was far more fascinating than if she'd been telling stories about a chimpanzee called "DG482." David Graybeard was an individual being with behaviors, expressions, and yes, even emotions.

Goodall wasn't alone in seeing nonhuman animals as individuals with unique personalities and emotional experiences; evolutionary biologist Marc Bekoff also named the animals that he worked with. Bekoff empathized with his animal subjects when he had to perform terminal procedures. He was so struck by the emotion of one of the cats that he worked with that after he euthanized the cat, he vowed to never harm another living being. Given the neurological and biochemical similarities between humans, chimpanzees, cats, and dogs, it's not surprising that those who work closely with them find much in common with them. But genetic similarity (e.g., genetic material shared in common between humans and chimpanzees) and co-evolution (e.g., the partnership between humans and domesticated dogs) doesn't yield identical emotional experiences between these species.

Psychologist Paul Ekman, who is perhaps most well known for his groundbreaking work examining microexpressions, studied emotions and catalogued more than 10,000 distinct facial expressions, of which 3,000 were relevant for emotions. Ekman expanded upon Darwin's theory that emotions were traits that evolved through their adaptive value and were universally shared across human populations. Across multiple cultural groups worldwide, Ekman found facial universality with shock, joy, loneliness, fright, grossness, and wrath.[104]

THE PHYSIOLOGY AND BIOLOGY OF EMOTION

In the human brain, the limbic system is the emotional control center. It's responsible for recognizing, processing, and controlling the emotions. Within the limbic system, there are several structures, including the hypothalamus, the amygdala, and the hippocampus. These are the "physiological" emotions that we share with other animals. Emotions like guilt, spite, and jealousy, while involving the physiological limbic system responses, also involve considerable cognitive processing, theory of mind (understanding what others are thinking or planning), and forethought. It is these "higher" or "cognitive" emotions that are probably only shared by our closest relatives, like chimps and gorillas, and perhaps a few other species with the highest forebrain development. While those who share their lives with dogs might feel that their dogs exhibit signs of these secondary emotions, there is no scientific evidence that dogs experience

---

[104] http://www.paulekman.com/wp-content/uploads/2013/07/An-Argument-For-Basic-Emotions.pdf.

them. While the pendulum has swung from emotional denial to emotional acceptance, well-designed research in emotions in animals remains hard to come by.

So where do we draw the line between the emotional experiences of humans and nonhuman animals? How do physiology and cognition interact? We can examine biochemical changes in nonhuman animals, but we still know very little about their emotions.

To measure emotion, we can quantify neurophysiological measures of positive emotional effects, including oxytocin, prolactin, b-endorphin, b-phenylethylamine, and dopamine, among both dogs and humans. And while most research to date has focused on the positive effects that dogs have on humans (including hormonal changes and decreases in blood pressure and heart rate), studies published in *Science*[105] and *The Veterinary Journal*[106] examined the reciprocal effects of these changes. This is at the heart of the examination of human-canine coevolution!

If dogs have a positive effect on humans, then do humans also have a positive effect on dogs? The answer, on the whole, is yes. The study notes: "Our results indicate that concentrations of (positive agents) b-endorphin, oxytocin, prolactin, b-phenylethylamine, and dopamine increased in both species after positive interspecies interaction, while that of (the negative agent) cortisol decreased in the humans only." The interactions mutually raised positive measures in both, but only decreased the negative measure in humans. This raises further questions: what's the real function or role of cortisol? This also makes us take more caution in interpreting cortisol's use in stress studies. Is it possible that mutual interactions may be more stressful to dogs than to humans? This is an important factor of recent concern in therapy dog use.

There are other ways to quantify emotions among dogs, including behavioral studies examining equity. For instance, there's evidence that dogs are capable of determining whether a human handler has acted "fairly," as defined by whether they received the same reward for a behavior that another dog did. For example, a dog who wasn't receiving a reward for giving his paw would begin to hesitate to do so if he saw that one of his peers received a reward for doing that same behavior. Similar results have been found in studies with monkeys and with human children. But what about their emotional responses to these inequities? Do they feel offended, frustrated, outraged, and confused when presented with situations that would elicit these emotions from humans?

While we still struggle to understand the emotions of our dogs, they continue to excel at reading ours. While we like to think that we're emotionally aligned with our significant others, humans on the whole actually aren't good at picking up on when they're feeling sad, but our dogs are particularly good at this. Our significant others are good at detecting when we're happy, but not so much when it comes to the darker emotions. We often assume that our partner feels the same way that we do, and the

---

[105] Nagasawa, M., Mitsui, S., En, S., Ohtani, N., Ohta, M., Sakuma, Y., et al. (2005). Oxytocin-gaze positive loop and the coevolution of human-dog bonds. *Science*, *348*(6232), 333–336.

[106] Odendaal, J. S., & 7 Meintjes, R. A. (May 2003). Neurophysical correlates of affiliative behaviour between humans and dogs. *The Veterinary Journal*, *165*(3), 296–301.

bias of our emotions interferes with our ability to detect theirs. Our dogs, however, do not suffer from this same bias.

While many people believe that their dogs can feel "guilty," this is highly debatable. That "guilt" is likely the anticipation of our anger or punishment, rather than their own feelings of shame or regret over a perceived transgression. Researcher Alexandra Horowitz found that dogs displayed specific behaviors (including avoidance of eye contact, downcast head, ears back, walking away in avoidance) in response to their owners' criticisms. Researchers determined that these behaviors weren't evidence of "guilt," but rather anticipation of punishment.

MAKING THE CASE FOR CANINE DEPRESSION

Only one century ago, most domesticated dogs were "working" animals with vocations that were enriching and kept them occupied. Many dogs were bred specifically for certain tasks, and they were happiest when performing those jobs. These dogs were bred for generations for a specific purpose, but in a relatively short amount of time, they were just "pets" – relegated to staying in the home while their humans were away at work. While technology can change rapidly, evolutionary change – even with the help of artificial selection – isn't as fast, and as Canine 2.0 has yet to make it onto the modern technological scene, many dogs are understandably suffering from the same mental issues that humans do.

Depression is a very real, biological, behavioral, and physiological response to the loss of a connection in humans, nonhuman primates, and virtually all animals in which it is has been researched. That "connection" might be physical, to an object, or a place, but frequently, it is to a social partner where a social bond has been broken. There's a loss, you're feeling insecure, and your endocrine levels alter your behavior. Many animals can and do experience depression, but can nonhuman animals experience "grief" as well? Is there a line between depression and grief? According to Anthropologist Dr. Barbara J. King of the College of William and Mary, author of *When Animals Grieve*, grief "requires that an animal's normal behavior routine is significantly altered, and that she shows visible emotional distress through body language, vocalizations, social withdrawal, and/or failure to eat or sleep. So, while I do need to follow consistent criteria in describing grief, I don't need to know what an animal is thinking, any more than I'd need to know what a person is thinking, if he shows marked emotional response to a death." Demonstrations of grief in nonhuman animals reveals the importance of studying their emotional lives to understand their behaviors and to improve their welfare.

Joy or elation, which are on the opposite end of the emotional continuum from depression, involve the release of endorphins, and these are fundamental hormones that are found throughout the animal kingdom. Is a dog happy when he or she gets out of the house to go on a walk? If they look like they are, then they probably are! We can base this upon our prior experiences with dogs – and with specific dogs – and with our knowledge from ethograms that can help us deduce, from ear to tail, whether the dog is likely having an affinitive or an agonistic reaction. Of course, for some dogs, leaving the house is terrifying, another emotion with a clear biological basis.

Yes, nonhuman animals experience emotions, but theirs are "basic" in comparison to the ones that humans experience. Human emotions often draw upon past experiences, observational learning, intuition, and higher cognitive function, which fundamentally requires Neocortex development – including intuition and planning – and the ability to see the future. These emotions are the ones that most species of nonhuman animals don't have – they simply don't have the brain structure to foresee in this way. But that doesn't prevent many pet owners from seeing what they want to see. When Alec and Ziva were walking through the woods, he perceived her expression to be "hopeful" and "expectant" after she had received two treats. But was she actually feeling "hopeful" that she would receive another treat, or had she learned that looking at Alec in that way elicited the desired response?

We share our lives with our animals, but misinterpreting our animals' emotional states is a common occurrence. A dog named Spike was said to engage in "spiteful" behavior when he took another dog's ball and hid it. When the canine owner of the ball came looking for it, Spike quickly walked off in the other direction, seemingly like a killdeer bird might walk away from a nest with eggs in it: to draw attention from the desirable object or area. But was Spike being "spiteful," or was he exhibiting some fundamental neuro-endocrine response? The ball was a desired resource; taking it from the other dog could easily be classified as "resource guarding." And walking away from the desired object was simply a way to distract attention from the hiding place. Did this involve "spite" at some cognitive level? There is no real reason to believe so.

This is almost always the case in JCH's dog in-home dog cases. There have been two unquestionable cases of depression in dogs, one in cats, and several cases in parrots. One of these cases involved two Australian Shepherds named Roscoe and Maddie. These two grew up together in a household, with Maddie arriving about two years after Roscoe, after the loss of another, earlier dog in the family. But at the age of 10, Maddie passed away from a degenerative condition. Roscoe, who was about two years older than Maddie, shut down. He ate minimally and played reluctantly or simply not at all. He began occasionally urinating in the house, which was the catalyst for my appointment with his family. This began to occur about two months after Maddie's loss. He appeared more like he wasn't feeling like going outside rather than intentionally urinating in the house. The owners began to suspect physical ailments and had Roscoe in to the vet for a thorough check-up, but he had a clean bill of health. The owners were then referred to JCH. I (JCH) assessed the situation, and immediately diagnosed depression due to the recent loss of strong social attachment.

More challenging is the treatment with a case like this: time is the best medicine in these cases. Anti-depression medications are available, but they take so much time to determine the effective type and dosage that, in most cases, they are best reserved for use in cases of "profound depression," which involves a true mis-balance in brain neurotransmitter chemicals and is usually long-lasting. In the case of Roscoe, it appeared that the depression would correct itself, and that the best medicine was supportive care.

His people needed to make sure that they provided him with enough to eat (such as high-value foods, if need be), they needed to walk him often, but usually briefly, and they needed to provide him with a lot of replacement social attachment (but let them tell you how much). If you've ever felt sad or depressed, this treatment plan might sound very similar to what worked for you.

There have been similar cases in cats, including a cat who lost their long-time owner. It took years for this cat to recover from his loss. JCH reports that he has also seen this with parrots, with the loss of either another parrot social partner, or a highly interactive owner. One case was clear-cut: the parrot demonstrated a lack of eating and severe feather-plucking occurred when the bird's beloved teenage owner left for college, and these behaviors disappeared during every return trip home. The interview took an hour to manage to confirm this pattern, but fortunately in this case, replacement attention (moving the parrot's cage to a busier part of the house) worked well!

The depression responses of the highly intertwined nervous and hormone system are themselves interlinked with all systems in the body. The saddest circumstances, as in humans, are of profound depression cases which have produced alterations in the neurohormone systems and which have manifested into systems that influence the immune system, eating systems, and even cardiac systems. The loss of a strong social attachment triggers deep and profound (especially so if it goes untreated) depression, which leads to other medical issues, and sometimes even death. JCH had a case once with a mixed breed dog named Alex who was deeply attached to his 78-year-old owner. When his owner passed away, Alex went deeply into depression. The owner's kids took in Alex, but it was not the same, and their vet brought him into the situation as Alex began to eat less and less, and to critically lose weight. He tried everything in the book, but it was not to be… Alex basically starved himself to death, and he fully feels that this was a case of profound depression. We hear about cases like this in the social media these days, and we believe most of them!

Just like humans, dogs can exhibit varying degrees of depression, from mild to profound. During a road trip from Washington State to Southern California, Jack had spent all of his time with TLC for an entire week. Driving, eating meals, exploring towns large and small, hiking, sleeping – every moment was spent with her. And during this trip, Jack likely had elevated levels of those "positive" hormones, oxytocin, b-endorphin, prolactin, b-phenylethylamine, and dopamine for a prolonged period of time. Jack exhibited higher energy levels than he usually did, even walking and hiking for more than nine miles one day (which is a lot for a tripod dog whose prior record was seven miles in a day). He was demonstrating frequent "dog laughs," sleeping less than he typically did, initiating play often, and his overall demeanor appeared to be happy. When the trip was over, though, life returned to "normal" and Jack would have spans of time several hours long where TLC had other commitments. Jack began to exhibit signs of depression: he became disengaged, quiet, appeared to be less interested in playing and solicited play less, had a decreased appetite, and had far lower energy levels than what was "normal" for him for several weeks after the trip (Fig. 27).

FIG. 27

Jack, TLC's black Labrador retriever mix. *Photographs by Sarah Bous-Leslie.*

THE DOGS OF GROUND ZERO

During the attacks on September 11, 2001, 2,753 people died in New York City, including 2,606 in the World Trade Center's towers and 147 people on the planes. In the hours and days following this tragedy, some 300 search and rescue dogs were deployed to search the ruins for signs of living victims. As the hours turned to days, and then to weeks, and the dogs continued to be unsuccessful in their search for living survivors, they began to exhibit clinical signs of depression. Without the reward of finding a living survivor, and accompanied by partners who were also distraught at not finding living victims, the dogs' motivation began to wane. Their handlers showered the dogs with affection during the course of their 12-hour shifts, but when their dogs continued to show signs of depression, some of the workers hid in the rubble to allow the dogs to find living humans in addition to the deceased victims.

Far from being anthropomorphic, recognizing that our dogs are susceptible to depression, just as we are, can help us better care for their physical and mental health. Dogs who exhibit signs of depression should be assessed for the root cause of this issue while taking into account what their breed was designed to do. A herding dog will easily become bored and then depressed in an environment without a lot of exercise and stimulation, and most dogs will become depressed over time without social interaction. Depressed dogs can be helped with rewards, redirection, more attention, and with enrichment items such as toys and food puzzles that can keep their minds occupied.

ARRESTED DEVELOPMENT

According to Dr. Temple Grandin, many of the emotional and behavioral issues that we see in our dogs today are artifacts of their modern urban lifestyles. Grandin notes that prior to the modern era, domesticated dogs were largely free-ranging on farms and ranches during the day; they weren't kept in kennels or even behind closed doors, but in yards without fences. As we have continued to move toward a more urbanized culture, though, our dogs have had to do so, as well. Our modern lifestyles limit dogs' natural abilities to roam and explore as they did only a few generations ago. As a result, they're exhibiting more and more unusual behaviors, such as stereotypies and other abnormal behaviors, including obsessive-compulsive disorder (OCD). Stereotypies are repetitive or ritualistic postures and movements that are often seen in captive animals, such as bears, wild felines, wild canines, and primates. Stereotypical behaviors can include pacing, rocking, over-grooming, and even self-injurious behavior, which can include over-licking or biting, resulting in damage to the underlying tissues.

While rare, some of the more common self-injurious behaviors include flank sucking, where a dog will repetitively suck on their own flank until they have denuded it, tail biting, an extreme form of tail chasing that can result in partial or complete tail amputation, and compulsive licking that results in acral lick dermatitis, where the flesh is ulcerated and infected. There might be an initial cause for these behaviors, such as pain from an injury, fleas, infection, or allergies, but the behaviors can become compulsive and, over time and without redirection, stereotypical.

Stereotypies among domesticated dogs, while fairly uncommon, include pacing and bouncing, and occur more frequently with kenneled dogs than with unkenneled ones. A study published in *Physiological Behavior* examined the repetitive behaviors and cortisol levels of 30 kenneled dogs in a variety of contexts, and in the presence of highly arousing stimuli, including food preparation, the presence of staff, and the presence of other dogs. Twenty-eight of the 30 dogs (93%) exhibited repetitive behaviors in some form, but only five of the dogs (17%) exhibited repetitive behaviors in the absence of these arousing stimuli. The majority of these repetitive behaviors occurred during the preparation of food or when a dog and a person walked by the kenneled dog, indicating that most of these reactions were likely temporary responses to high-arousal situations. Cortisol level analysis revealed surprising results, as well. All of the dogs had a similar baseline with their cortisol levels, but the cortisol levels of the five dogs who exhibited repetitive behaviors in the absence of stimuli had decreases in their levels post-exam, rather than the increase that was shown by the other dogs, potentially indicating signs of long-term stress. This would make sense if these dogs were exhibiting stereotypies.[107] Stereotypies are most frequently seen among dogs who live in environments with fairly consistent, high levels of stress, such as dogs that are used for laboratory research or dogs who are left tethered outside without frequent socializing.

Stereotypies and other indications of mental instability can also arise among dogs who are living in "ideal" conditions, but are exhibiting unusual reactions to stimuli. Some dogs might become ball or laser-obsessed, which appears humorous on the surface level, but actually indicates errors in how their brain is processing these stimuli.

---

[107] Denham, H. D., Bradshaw, J. W., & Rooney, N. J. (2014). Repetitive behaviour in kenneled domestic dog: Stereotypical or not? *Physiological Behavior, 128*, 288–294.

In JCH's in-home practice in Seattle, he has seen a couple of cases of obsessive-compulsive disorder in dogs, and very quickly handed them over to a qualified veterinary behavior specialist at Washington State University. He has had two other somewhat-related cases: the first to come to mind is a case of a dog which developed what he would describe as almost a panic disorder in the face of moving reflections and even shadows which was tied back to the use of a laser pointer for stimulation: Chase the Red Dot. The dog seemed to enjoy this stimulation, but within days, began to develop a reaction to moving reflections of all sorts, not quite acting scared or anxious but also over-excited, over-stimulated. He was knocking over furniture and lamps in his furious reaction to reflections. JCH had heard of this reaction to other dogs from some of his colleagues, so while it is little documented, it is not unknown in the business. He speculated that it is most likely some sort of error, either existing or newly-formed, in the brain that produces this reaction.

Another brain-wiring, neuroscience-based case was the case of two Boston terriers in Vancouver British Columbia whose owners used remote consultation: one dog would suddenly, for absolutely no apparent reason, attack and bite the other dog. The other dog was described as "just minding his own business" in every case. A camera was left turned on, on the floor in the evening when these incidents would happen every few weeks. Once we captured an incident and the video was posted to JCH, it was immediately noted that the target terrier had experienced a (mild but distinct) seizure in the seconds before the aggressor attacked. We waited another week or two, captured another occurrence on video, and confirmed the same pattern. The victim was carted off to a veterinarian who successfully treated the seizures and the behavior was never seen again. In each of these cases, a neurological disorder is at the root cause of the behavior.

Dogs were bred to be social animals, but as they are being left alone in the home for increasingly long periods of time, behavioral issues such as separation anxiety are arising. Dogs who are going to have separation anxiety typically present their behavioral issues, such as barking, whining, howling, pawing, chewing, and destroying objects, within the first hour of separation from their human. Certain breeds, such as German Shepherds, were bred to be particularly vigilant, and thus suffer from higher anxiety levels than dogs without that genetic heritage. Being aware of breed-specific and individual needs can help decrease the incidence of emotional issues with our dogs.

WHY IS EMOTION SO IMPORTANT?

Emotions are critically important to our communication: they enable us to understand one another, they're reflective of our internal endocrine state, and they help us relate to the experiences of one another. Smiles, laughter, yawns, and even crying can be "contagious" to a certain degree; if someone is telling a story and laughing as they recall it (but they haven't yet reached the "funny" portion of the story) the recipient might begin to laugh in anticipation, before they even hear the funny part. This demonstrates empathy, something that our dogs demonstrate with us, as well.

Have you ever yawned to your dog? Our dogs will often "catch" our yawns, just as a human recipient might. But until recently, there had been little research on the neuronal aspects of this phenomenon. Researchers in a 2013 study[108] hypothesized that the human mirror neuron system (MNS) is activated when an observer perceives another person yawning. Functional magnetic resonance images assessed brain activity during contagious yawns and subjects demonstrated activation of their Brodmann's Area, a region of the right inferior frontal gyrus, which is part of the MNS. The results of the study revealed that the initiator of the yawn and the observer who then "caught" the yawn shared both emotional and physiological states, thus demonstrating that contagious yawning is based on empathy. Given both the continuity of neural circuitry across mammal species and the incidence of contagious yawns from humans to dogs, it's valid to make the inference that dogs' mirror neuron systems are similarly activated by contagious yawning (Fig. 28).

The behaviors and emotional reactions of others affect us on a physiological level, but what happens when we perceive a mismatch between visual and verbal messages? During his tenure as President of the United States, George W. Bush had to deal with 9/11, the hunt for Osama bin Laden, and his own emotional mismatching: his visual emotional cues often didn't correspond with what he was saying. Bush often delivered bad news with what looked like a smirk on his face, but this was likely just an unfortunate case of nerves and not the world's worst case of schadenfreude, wherein one derives pleasure from another's misfortune. But what he was saying wasn't always well received, because the accompanying visual cues were muddying the content of his verbal message. And auditory and visual mismatches aren't just disconcerting to human audiences; our dogs pick up on this, too. While humans will criticize those who demonstrate communicative mismatches, our dogs are experts at detecting our true emotions and will act accordingly. For example, we

FIG. 28

Dogs yawning. *Photographs by Sarah Bous-Leslie.*

[108] Haker, et al. (2013). Mirror neuron activity during contagious yawning—An fMRI study. *Brain Imaging Behavior*, 7(1), 28–34.

might say that we aren't upset about something, but our dogs see beyond this, picking up on our subtle visual and chemical cues.

Whether they're picking up on pheromones, gestures, or tones of voice, dogs are masters of reading our emotional states – but for their optimum mental health, we need to tune in to their emotional lives, too. If a dog was depressed for a long period of time, as Alex was, or even for a short period of time, as Jack was, trying to train them during this time would be setting them up for failure. Similarly, trying to work with a dog when they're angry or frightened wouldn't yield favorable results, either.

Tuning in to a dog's behavioral indications of their emotional states is important. If a dog is exhibiting behaviors such as licking their lips, avoiding eye contact, tucking their tail, and turning their head away, this isn't the time to intrude into their space or try to teach them something new; they're communicating, as best as they can, that they are stressed and uncomfortable. Ignoring these signals can push a dog into a further state of discomfort, resulting in escalated actions, including growling and biting. Far from being "automata," dogs are emotional beings who express their feelings in myriad ways, much like their human companions do.

An important topic, and the trigger for much of dog-dog and even dog-human "misunderstandings" and possible aggression, is the inappropriate *communication* of emotions. Now, communication is a broad topic, and a great deal of information can be communicated between individuals, but for this purpose, we're referring to the communication specifically of the emotional state. As primates, and higher-cognitive functioning ones at that, we know well how important emotions are in the life of the individual, and how important it is to correctly perceive the emotional state of other individuals with whom we interact. But how well can we determine the emotional state of other species? And imagine the confusion across other species with less cognitive function than we have!

One of the most important factors in misunderstandings between dogs, which we see in the dog parks and often results in aggressive behavior, may be the lack of ability to appropriately read body language and communication between those individuals. This communication is important because it transmits the other dog's emotional state, and this implies a lot about intention.

Another point to remember is that as primates, we are very visually-based and so we are focused on body language. But remember that the dog may be communicating in different dimensions. Emotional state might be transmitted by pheromones, chemicals given off by the body, or it might be transmitted by high-frequency or very-low-frequency vocalizations which we can't hear, or certainly tend not to perceive, and so we miss these messages entirely.

So the changes that we've made in breeds of dogs which have altered their physical appearance have probably altered their ability to communicate these emotional states amongst each other and quite possibly with humans as well. So to some extent, we have handicapped ourselves, and our dogs, by altering their physical emotion-communication capabilities to the point where we may be beginning to lose some of the tight co-evolutionary relationship between humans and dogs, and generated dogs which may have trouble with social relationships with other dogs.

Why do so many people ignore the emotional state of their dogs? We can open any page on Facebook or look at any website on dogs, and people interacting with their dogs; we can watch any dog training or behavior show on television, and see dogs that clearly in my opinion are displaying an emotional state contrary to what the human owners are applying to that pet.

Usually this involves a dog which is fearful. Why do so many people ignore the fear and anxiety signals of their dog? Is it that they simply don't have the knowledge to recognize fear and anxiety signals in their dogs? We are not sure that this is true and we often think that people, when asked, will tell you that their dog looks anxious but for some reason, they seem to think that it's OK for their dog to be anxious in a particular situation. To this day, we still fail to understand how this might be true! We don't think that these owners are necessarily bad people; we think it's more likely that they have trouble believing that dogs have these fundamental emotions, or maybe don't understand how fear and anxiety might have a dog feel. Of course, presumably the owner understands how it feels to be fearful or anxious but there is some disconnect to being able to extend that understanding to another animal.

Now, on the other hand, we have the cases of extreme anthropomorphism in which owners are extending complete and high-level human emotional states to their dogs, including jealousy and spite and love and other higher-level human emotional constructs. In a similar fashion, the tendency for this behavior complicates the relationship between human and dog just as much.

These tendencies are a factor in clinical cases: most of our behavior modification methods rely on the handler being able to recognize the subtle, early-warning signs of anxiety and performing some sort of reward. But when the owners *cannot* recognize the signs of anxiety, it doesn't work. Much of JCH's time, or his trainer's time, is spent in training the ability to read the signs, before we can even move on to what to do about them!

And likewise, the over-anthropomorphism results in just as many roadblocks: it's *not* a baby, it's certainly not *your* baby, it's more capable than a baby, and even babies need rules, structures, and training (okay, we usually call it "learning"!). JCH's work becomes the need to gently disabuse the owner of these concepts, and to persuade them that their "baby" would be happier with some rules, or some structure, or some training!

CHAPTER 6

Is it worth the risk: How costs and benefits drive decision-making and the evolution of behavior

"RISKY" BUSINESS?

Jack eyed TLC intently as she opened a can of cat food. Jack, who's a medium-sized dog at about 60 pounds, is an "average" domesticated dog: he loves to play, go for walks and car drives, and, like most dogs, he loves to sneak things that he's not supposed to eat. He watched as his owner filled one bowl with soft cat food and then approached a second bowl. As she turned her back, he made a beeline for the first bowl, devouring its contents in large, chewless gulps. The cat food was gone before she turned around.

"Jack! No!" TLC yelled, "It might make you sick!" and it did. Hours later, Jack was vomiting and shivering. Despite her best efforts to deter him, Jack has gotten into the cat food several times, and given the chance, he'd sneak the cat food again tomorrow. But why, when it sometimes makes him sick, and he's yelled at when he's caught eating it, does he continue to do this? Because he isn't caught (or yelled at) during every instance of food stealing, and he becomes sick from the food just infrequently enough that it's worth the risk. For him, the "benefit" of eating the soft, pungent cat food outweighs the "cost" of potentially getting scolded or getting sick. The risk is worth the reward. If Jack were to become sick each time that he ate the food, he would develop a taste aversion, a phenomenon wherein nausea works in the brain to produce an avoidance of certain foods. Taste aversion is an adaptive trait that trains the organism to avoid potentially poisonous substances; this survival mechanism is an example of Pavlovian conditioning (also referred to as classical conditioning).

Pavlovian conditioning, named after Russian researcher Ivan Pavlov, involves pairing a biologically potent stimulus such as food with a neutral stimulus. In the early 1900s, Pavlov worked with dogs and paired a bell with food. Over numerous experimental trials, the dogs in his study began to exhibit anticipatory salivation when they heard the bell, irrespective of the presence of food. But if a behavior is intermittently reinforced, as is the case here, and Jack only has an adverse reinforcement (such as scolding and/or sickness) a fraction of the time, the behavior remains worth that risk.

Humans experience a similar phenomenon when they play the lottery: one ticket may be a small risk at only $1, but the potential winnings are in the hundreds, thousands, or even millions of dollars. Most tickets are non-winners, but intermittently winning $5 or $10 is just enough reinforcement to perpetuate the behavior.

Intermittent reinforcement is a conditioning schedule where punishments or rewards are inconsistently administered. This is in contrast to continuous reinforcement, wherein a behavior is reinforced each time. With gambling, whether it's the lotto, cards, or slots, there isn't a payoff each time, but when there *is* a reward, it's a powerful enough reinforcement to compel the individual to continue to gamble. The possibility of this reward is so powerful, in fact, that gambling addiction has been added to the *Diagnostic and Statistical Manual of Mental Disorders, 5th Edition (DSM-5)*. It's risk-taking gone awry. In moderation, though, risk-taking behaviors (seeking out a mate, seeking out better food resources) can help perpetuate the genes of a species, and scientists have discovered that there's a neurological basis for risk-taking behaviors.

Risk-taking triggers the brain's reward system. The reward system has a region in the basal forebrain called the nucleus accumbens, or NAc or NAcc, which contains dopamine receptor cells called DR1 and DR2. In a risk-taking experiment published in the journal *Nature*,[109] researchers tested whether rats would convert from risk-averse tendencies to risk-preferring ones. The rats in this experiment played a game wherein they could trigger one of two levers: the first producing the same amount of sugar water each time and the second regularly yielding a small amount of sugar water and irregularly yielding a much larger payoff. The rats were given the opportunity to play this gambling game 200 times per day, with two-thirds of the rats pushing the reliable sugar water lever and one-third seeking out the sugar water lever with the potentially higher (but inconsistent) payoff. The researchers observed that the DR2 cell activity spiked for the conservative one-third of the rats whenever they chose a lever. When the researchers gave the rats a small dose of pramipexole, a drug that's been correlated to impulsive gambling behaviors in humans, the conservative rats deserted their risk-averse tendencies and began to touch the "high-risk" lever. The experimenters were thus able to convert the risk-taking responses of the rats by precisely timing the stimulation of their DR2 cells.

Gambling, like drug use, is associated with an adrenaline rush, but these aren't the only behaviors that are associated with risk-taking. From high-stakes choices like gambling to mundane daily activities, humans make similar decisions, using cost-benefit analyses multiple times every day, often without realizing it. From the moment we wake up (do we hit the snooze button again for nine more minutes of sleep, with the cost of having to rush to get out the door in time?) to the moment we go to bed (do we stay up half an hour longer to read one more chapter or watch one more episode of our favorite program, with the cost of being more tired in the morning?), we're weighing out the costs and benefits of our choices. Will we speed so we aren't late to work or an appointment, risking the possibility of receiving a ticket that's both financially and temporally costly? Will someone taking a test cheat in order to get a higher score, even though getting caught cheating is very "costly" and could possibly result in a score of zero? We face multiple decisions throughout the day: break the rules or don't, cheat or don't, make a white lie or don't, and we're constantly

---

[109] https://www.nature.com/nature/journal/v531/n7596/full/nature17400.html.

weighing the pros and cons of each of these decisions. Big risks often yield big rewards, but they can yield big costs, too. This is evidenced by the occupancy rate in jails across the United States – more than 2.2 million people, as of this writing, who gambled that the risk of illegal behavior was worth the potential reward.

RULES OF THUMB

Whether it's financial, emotional, or biological, no one wants to invest unwisely. As a general rule, an animal will be more likely to perseverate in a behavior if it's a wise investment. This is basic cost-benefit analysis, a systematic approach that can be understood through both proximate and ultimate explanations. Proximate explanations concern the developmental, neural, and endocrine events within an individual's lifetime that reinforce their behavior (the "how"), while ultimate explanations concern the behavior's fitness or evolutionary consequences (the "why").[110] Ultimate explanations arise through natural selection shaping the proximate mechanisms, and thus the behaviors, of individuals over many generations. Evolutionary strategies like kin selection and reciprocal altruism both exemplify cost-benefit analysis, one for biologically related members of one species, usually within a social group, and one for animals that don't have to be related (or even belong to the same species). Reciprocal altruism, a concept initially developed by biologist Robert Trivers, is evolutionary biology's version of "I'll scratch your back if you scratch mine."

Reciprocal altruism has been exhibited by multiple species, including animals that participate in cleaning symbiosis, reciprocal grooming among primates, and arboreal species (including birds and monkeys) making alarm calls to warn other individuals about the presence of predators. While evolutionary explanations for reciprocal altruism have been thoroughly explored, there's little in the scientific literature detailing the proximate mechanisms of reciprocity.

An animal will behave in a manner that might temporarily reduce his or her fitness while increasing another's fitness, with the expectation that this favor will be reciprocated. Consider, for example, domestication: an animal might acquiesce to pulling a cart, expending their own energy for the benefit of another, in exchange for later receiving food, water, and shelter that might have been harder to come by on their own. The cost of apparently altruistic behaviors in terms of reduced evolutionary fitness occurs because of the risks involved when an animal behaves heroically, as in when one member of a social group rescues an unrelated member from a predator, but this can be offset by the likelihood of the return benefit at a later time.

How does cost-benefit analysis present in our daily lives? According to Duke psychologist and behavior economics professor Dan Ariely, cheating is a prime example of cost-benefit analysis – and it can be viewed in economic terms as well as biological ones. Ariely asked, "…what's the probability of being caught? How much do I stand to gain from cheating? And how much punishment would I get if I get

---

[110] Tinbergen, N. (1963). On the aims and methods of ethology. *Zeitschrift für Tierpsychologie*, 20, 410–433.

caught? You weigh these options out – you do the simple cost-benefit analysis, and you decide whether it's worthwhile to commit the crime or not."[111] No one who ever steals, lies, or cheats when they think that they're going to be caught; the benefits of these choices have to outweigh the costs.

Humans aren't the only animals that use cost-benefit analysis in their decision-making, though; it's also a fundamental principle in nonhuman animal behavior. If a behavior is adaptive, the benefits will outweigh the costs. If it's not, then the costs will outweigh the benefits. And why would an animal continue a potentially detrimental behavior? On a proximate level, animals will take into consideration costs such as caloric expenditure and temporal investments, weighing them against benefits such as protection from predators, thermoregulatory support, and an increased availability to resources. On an evolutionary level, animals make choices that will result in maximizing their potential to perpetuate their genes into future generations. While animals don't necessarily consciously calculate a behavior's costs and benefits, animals that behave in a manner consistent with maximizing benefits relative to costs are more reproductively successful and experience higher levels of fitness,[112] thus perpetuating their genes across multiple generations.

Niko Tinbergen, who was the first to present animal behavior in terms of costs and benefits, applied cost-benefit analysis to black-headed gulls' (*Larus ridibundus*) removal of broken eggshells from their nests.[113] Tinbergen conjectured that the broken eggshells were conspicuous and attracted predators; the gulls removed the shells to decrease the risk of predation. Animals have to weigh the costs and benefits of time-consuming behaviors that detract from their ability to participate in other activities, such as foraging and hunting. The temporal and energetic costs of removing the shells was worth the benefit of added protection from predators and the chicks' increased potential for survival.

SISTER ACT

As much as we refer to ourselves as our "pet parents," kin selection doesn't apply to the human-canine relationship, but it can explain *intra*-species behaviors. Scientists have tried to quantify kin selection for decades – and none summarized it more accurately than J. B. S. Haldane,[114] whose relational synopsis predated W. D. Hamilton's more famous "Hamilton Rule" by three-plus decades. In 1932, when asked if he would give his life for a brother, Haldane quipped, "I would lay down my life for two brothers or eight cousins." Across animal species, it looks like there's some truth to that.

---

[111] http://www.ted.com/talks/dan_ariely_on_our_buggy_moral_code/transcript?language=en.
[112] Ha, R. R. (2010). Cost–benefit analysis. In M. D. Breed & J. Moore (Eds.), *Encyclopedia of animal behavior, Vol. 1* (pp. 402–405). Oxford: Academic Press.
[113] Tinbergen, N. (1953) *The Herring Gulls World. New Naturalist Series*. London: Collins.
[114] Haldane's study included mathematics, evolutionary biology, and genetics.

So what are some of the most important factors in animals' cost-benefit analyses? Kin selection, or "inclusive fitness," plays a big role. Inclusive fitness marks an individual's ability to pass its genes on to succeeding generations while taking into account genes that are also passed down by their close genetic relatives. Not all animals within a social group will have their own offspring, but they will be motivated to support the offspring of their relatives, under certain circumstances. According to Hamilton's Rule, which is a central component of inclusive fitness, animals are typically more likely to help those that they are closely related to than those they are not. Hamilton's Rule is defined as $rb - c > 0$, where r refers to the genetic relatedness of the recipient to the individual, b refers to the additional reproductive benefit that the individual will receive, and c refers to the cost incurred for the individual. In other words, the benefit gained by an act, relative to its cost, is discounted by the distance of the relationship between actors, but the discounted benefit must still outweigh the cost for a behavior to remain. Or, the benefit is greatest when actors are closely related; behaviors between close relatives are going to be the most common.

HAMILTON

Long before the wildly popular musical about Founding Father Alexander Hamilton and the social climate during and after the Revolutionary War burst onto the scene, Hamilton's Rule explained other social phenomena. Hamilton's Rule was officially introduced to the world in 1964 when W. D. Hamilton wrote about kin selection, but Darwin alluded to this concept decades before in the *Origin of Species*, wherein he discussed the lives of eusocial insects (Hymenoptera) that were social but sterile. Hamilton hypothesized that Hymenoptera exhibited eusociality because this species has a trait of haplopoidy, or sex-determined reproduction. Eusociality, which is exhibited among some species of crustaceans, insects, and mammals, is defined by the cooperative care of the offspring of other individuals, with labor segregated into reproductive and non-reproductive groups. Not all honeybees reproduce, but they live in family groups and their support of the group as a whole helps perpetuate the gene pool of the family. Haplopoidy occurs when an organism only has one set of chromosomes, half of the typical diploid chromosomes. The males of these species are produced from parthenogenesis: they come from unfertilized eggs and are haploid. The females are diploid and develop from fertilized eggs, thus, sisters share 75% of their genes in common with one another, while mothers and their offspring only share 50% of their genes with each other. From a genetic standpoint, then, it makes more sense for sisters in this species to support their mothers in raising more sisters rather than raising their own daughters (Fig. 29).

Cost-benefit analysis doesn't just apply directly to reproduction; it also applies to life strategies that will enhance one's quality of life (and indirectly, to reproductive success.) Nonhuman animals, like humans, make numerous choices relative to costs and benefits. For many species, domestication was one of those choices—and with humans and canines, there is a long-established relationship between domestication

FIG. 29

It makes more sense for Hymenoptera sisters to support their mothers in raising more sisters than raising their own daughters. *Sketch by Arianne Taylor.*

and reciprocal altruism. According to Jared Diamond, author of *The Red Queen* and *Guns, Germs, and Steel*, the domestication of certain plants and animals was the most important development for humans over the course of the past 13,000 years. Nowhere are the effects of domestication more evident than among modern domesticated dogs, whose variety would confound inexperienced observers. Diamond wrote: "What naïve zoologist glancing at Wolfhounds, Terriers, Greyhounds, Mexican Hairless Dogs, and Chihuahuas would even guess them to belong to the same species?"[115] While the ancestors of each of these dog breeds were selectively chosen over generations to produce the variety of animals that we have today, the dogs who were chosen to share their lives with humans were doing some choosing of their own: was it "worth it" to them to live with another species? If so, why? If not, why not? Over the generations, the dogs who decided that the costs outweighed the benefits either didn't join humans in the first place or, once living among humans, were culled when their temperaments proved to be unfit for cohabitation.

Across animal species, an individual can be said to be "risk-averse" or "risk prone," just as the rats in the sugar water study demonstrated. Those who are risk-averse may delay taking action with over planning and miss an opportunity, such as finding a suitable mate or moving to a habitat with richer resources, while those who

---

[115] Diamond, J. (2002). Evolution, consequences and future of plant and animal domestication. *Nature, 418*, 702.

are risk takers may take swift action with too little planning, potentially becoming injured or even dying. The risk-averse rats in the sugar water study received a reliable sugar water salary with their strategies, while the risk prone rats received either a smaller amount of sugar water or a highly prized sugar water lottery. These are differential strategies that are demonstrated across the animal kingdom. Among hunter-gatherer societies, those who are risk-averse focus on reliable but unexciting food sources, such as roots and nuts, while risk prone members seek out food sources like honey, which involves ascending trees and subsequently has a high mortality rate; or hunting big game, which runs the risk of injuries or death. Honey and big game are high-risk, high-reward items that are highly esteemed among hunter-gatherers, yielding both caloric and social benefits, but not all members of a social group find it "worth it" to participate in these higher risk behaviors; like the conservative rats, they'd rather pursue the reliable resource. Canines, too, exhibit differences in their risk taking rates across breeds and species. Domesticated dogs and wolves last shared a common ancestor less than 50,000 years ago and a recent study found that wolves are much more risk prone than domesticated dogs.[116]

THE SECRET LIFE OF PETS

Humans and dogs have co-evolved for millennia and the domestication of dogs has yielded costs and benefits for both species. Humans have the added cost of their food, but the added benefit of their protection; the added cost of their care, but the added benefit of their companionship. Over the generations, humans artificially selected those dogs who didn't guard their resources too strongly; the human-friendlier animals were selected over generations, being artificially selected to breed with other human-friendly dogs. Across multiple populations worldwide, the benefits for both species outweighed the costs. Dogs in diverse geographic areas found jobs as sled pullers, hunters, guardians, pest deterrents, and companions. The ancestors of dogs like Ziva chose to live with people, forsaking the relative freedom of hunting and choosing when they would pursue their prey for the relative security of regular but human-controlled meals. Over the generations, dogs learned how to influence and even manipulate their human partners, discovering which behaviors elicited the desired result. Young or submissive dogs and wolves licked the mouths of the "parent" members of their family group; this type of behavior transferred over into begging from human "parents." The ancestors of dogs chose security, consistency, and human companionship over their previous freedoms, and the dual wager paid off for both species. The wolves that Ziva saw were bedraggled and expending large amounts of energy to hunt, remain warm, and ward off rival predators. Ziva did not have to expend the same amount of energy. She benefited from the protection, warmth, and consistency of her human family. Thus, humans choosing to cohabitate with dogs and dogs choosing to cohabitate with humans was adaptive for both species, in both proximate

---
[116] http://journal.frontiersin.org/article/10.3389/fpsyg.2016.01241/full.

(time and energy) and evolutionary (increased reproductive success for both species) terms. The ultimate goal for every species is perpetuating their genes in future generations, and mutually beneficial partnerships are one route to accomplish this.

We weigh costs and benefits every day, so consider your dog's behavior in the same way. How is your dog being rewarded or inadvertently being punished? When your dog barks at someone coming to the door, how is that behavior being rewarded? Costs for such a behavior include their person yelling, "No!," but does the dog still feel that they're helping their people by alerting them to a possible danger? Jack the Lab mix will refrain from drinking the water that's in his kennel, even though it's readily available. In his kennel, he can't go to the bathroom, and will refrain from drinking even if he's thirsty. Being thirsty is worth it to him, though, as the immediate reward of remaining dry and not soiling his home outweighs the cost of temporary thirst.

With our dogs, viewing costs and benefits in a practical, immediate way might be more apt; we could even term the analysis "cost and reward" rather than cost-benefit when it pertains to canines. Successful dog trainers know that finding a dog's motivation is the key to dog training; so, too, do owners determine how to get their dogs to continue or discontinue certain behaviors.

A dog's perception of the world is different than ours; we rely heavily upon visual cues, but they rely primarily upon olfactory ones. How does this impact their cost-benefit analysis? When we're choosing a residence, we might be influenced by the fact that it's a view home, whereas a dog would be impressed by a home's wide array of odors. (Not that our dogs are choosing our homes, but you get the point.) Their sensory priorities are different from our own, but their needs are the same. During the domestication process, they likely smelled humans cooking food; despite the presence of fire, that odiferous reward was just enough to tempt some dogs to come closer to humans. Something that could get dogs more food, and with lower relative costs, would eventually help dogs get more of their genes into future generations, so there were both immediate and evolutionary benefits to these choices. Losing some of their freedom (such as when they received food, and what food they received; where they could and couldn't go, if they were on a leash or hooked up to a sled), these were all costs that they weighed and determined to be smaller than the benefits.

While humans and canines have co-evolved for millennia, they are not our "kin," and Hamilton's Rule does not apply, but reciprocal altruism is an apt model for explaining their behavior. According to this model, the dog thinks, "I'll help you out in this situation, and then you'll help me out later." While the dog might not realize that a behavior such as begging for food from humans will likely help them get more of their genes into future generations, they will understand the proximate benefit. When we are studying a certain behavior, we still need to keep in mind that the ultimate goal is to get more copies of one's genes into future generations.

Young dogs treat familiar dogs much differently than they do unfamiliar ones. When Ziva met the unfamiliar dog in the woods, she was hesitant; hypothetically, the dogs who are around them when they're younger would be related to them, while unfamiliar dogs wouldn't be. Even if they aren't close genetic relatives, young dogs treat familiar dogs in this way; puppies are impressionable. Adult dogs don't assume that other dogs are relatives and have attachment issues in comparison to younger dogs. That's why the early socialization window is so crucial – between 8 and 16 weeks of age, it's important for a puppy to socialize as much as possible. The puppy knows that their parents will keep them away from anyone unsafe. Humans have learned how to game that system – by understanding it, they can help their dogs remain more motivated. You have to look at your dog's behavior in terms of both how they perceive the world and with costs and benefits in mind.

Let's revisit the "guilty" example. While we anthropomorphize that our dogs "feel bad" about a perceived transgression, what we are actually observing is their anticipation to our displeasure at their "risky" behavior. The dog wonders: should I do this behavior? Is it "worth" the risk of getting caught? Similarly, we might misinterpret our dogs' reaction to a novel place or object. With something like an agility tunnel, a dog might not want to go in (and might be afraid of the enclosed space), but they realize that their owner is happy when they do so, and it's also paired with a food reward afterward. The benefits of this activity would likely outweigh the costs for a lot of praise and treat-motivated dogs. Owners should also re-assess the costs and benefits of punishment techniques, as the cost often outweighs the benefit.

Dogs have that close co-evolutionary history with us, while cats have shared our lives for a far briefer amount of time, evolutionarily speaking. Forty thousand (or more) years of human-canine cohabitation versus 9,000 years of human-feline cohabitation is a significant difference. Dogs know, "If I do this, my owner will praise me, which I like." This reinforcement doesn't necessarily mean as much to a cat as it does to a dog.

To comprehend a dog's behavior and why they make the cost-benefit choices that they do, you have to try to view these costs and benefits as your dog might perceive them. An ethological approach is at the heart of it. Consider visiting the vet, for example. Dogs experience the world through their umwelt, paired with prior experience. Your dog's perception of the veterinarian will be shaped primarily by their prior experiences and by the odors (the fear of other dogs, urine, medication, cleaning solutions, etc.) at the veterinary clinic. Many dogs will experience anxiety when they go to the vet, especially if they only visit when they're sick or need vaccines. Owners can counter condition their dogs by manipulating the cost-benefit scenario: for example, nine out of ten times, the dog only visits the vet's office to be weighed and get a treat; the tenth time, they receive a full check-up. Our dogs will come to view the vet's office differently when we manipulate costs and benefits in this way.

But learning principles work both ways: animals manipulate the behaviors of humans just as much as humans manipulate the behavior of animals. An animal can learn, "When I growl on the couch, the person moves away"; or "When I have this expression on my face, I receive a treat"; thus, the behaviors of growling or begging are reinforced. Animals can learn through both positive and negative learning, which operate on different brain pathways and use unique dopamine receptors. Revisiting the brain's reward center, the DR1 receptors are generally viewed as "positive" receptors, while DR2 are "negative" receptors. Across multiple species, studies have shown that individuals will prefer to stimulate their DR1 receptors over their DR2.

While punishment, which stimulates the DR2 receptors, might appear to have an immediate result, stimulating the DR1 receptors is more effective for learning in the longer term. Combining positive and negative with one another neurologically and psychologically confuses the learning process.

At some level, behaviors from an animal, such as growling or begging, and the recipient's response, merge into communication, but this can quickly get out of control. These issues can present in multiple ways. During his 30+ years as an animal behaviorist, JCH has amassed numerous case studies for the costs and benefits of canine social behavior. Just like with humans, domesticated dogs have individual differences in sociality.

A major way that to think about the relative costs and benefits of being social on a practical level is in the case of individual differences in sociality. The relative costs and benefits of being social for an individual dog frequently manifests itself in personality: for example, extroverted versus introverted, aggressive versus timid, high explore drive versus low.

Different stimuli are more salient not only to one species than they are to another, but also to different dog breeds and to individual canines. Dogs have been specifically bred for certain traits for so many generations that we have shaped their motivations just as much as water has carved the Grand Canyon, but in a fraction of the time. Not only do we have to consider our dogs' umwelts, and that their perceptions and priorities are distinctly different from our own, but we also have to take into consideration their individual personalities. An entire field of study has been devoted to "typing" humans by how they perceive and navigate the world: are they extroverted or introverted? Do they rely more on sense or intuition? Are they a "thinker" or a "feeler"? Do they rely more on judgment or on perception? Our dogs exhibit just as much individual variation. When we compound this with their species and breed differences, that's a lot of biology and psychology to sort out!

Before drawing conclusions about a behavior issue involving social interactions, it's important to assess how the *dog* balances the relative costs and benefits of being social. This can be done in the usual three ways: general knowledge of personality of the breed (if applicable), interview of the owner(s) ("How does Fido act in this (social) situation, or that situation?"), and my observations of the dog in social situations (and again, in situations in which the dog's problem behavior arises: off-leash park observations tell us little about leash-walking issues!)

Then, fundamentally, the behavior issue (usually aggression) should be viewed in the light of the relative costs and benefits…when the (potential) costs outweigh the benefits (in the dog's mind), then problems ensue. But exactly what are those (perceived) costs? What might the benefit(s) be? And how can we manipulate them to solve a problem?

In the case of Angus, a purebred toy poodle who is leash-aggressive towards other dogs, there were several considerations for JCH. He knew that toy poodles, as a breed, aren't highly (dog)-social, and he moved forward first by interviewing the owners about the problem behavior and when it occurred, specific situations, and then participating in a walk around a few other dogs, to see what happened.

Clearly, Angus liked his walks, exhibited eager anticipation behavior as harnesses and leashes were prepared, had a bounce in his step when they left the home, and remained in a positive mood… until we (he!) saw another dog. Distance was a factor, as it often is…mood remained positive while the other dog remained in the distance: the benefit of being on a walk outweighed the anxiety (or even perhaps the potential cost of being attacked, in Angus' mind). But as the other dog approached, Angus' body language revealed the changing equation, to the point that no amount of fun walk outweighed the (perceived) cost of that approaching dog.

JCH frequently explains the situation to owners in just these terms; it usually makes a lot of sense to them! And what to do about it? Bring out the high-quality treats: while we can institute a formal counterconditioning process, with below-threshold stimuli presentation associated with high-value rewards (otherwise known as, 'Associate the scary trigger with the happy food!'), but in many cases, without being so formal, he has had great success with what he believes is simply altering the perceived costs and benefits: the costs of an approaching dog begin to outweigh the benefits of being out on a walk (Sights! Smells! Owner-attention!), until the appearance of hot-dog treats swings the balance back again. Can we use associative learning to form a new conditioned emotional response (counterconditioning)? Sure… but very frequently, it is simply about the relative costs and benefits, a fundamental principle of modern animal behavior science!

JCH says that as another example, we can take his own dog, an Australian Shepherd named Elsie (a tribute to a founder of our science and early Woman in Science, Elsie Collias). She was *extremely* social, so the opportunity to socialize, the benefits of socializing, outweighed nearly all (perceived, or real) costs, like scaling fences, lunging and coming up short on a long leash, and stern reprimands from her handlers. The owners' clear disapproval meant nothing if there was another dog to play with in the (non-off-leash) park… and if that dog wagged its tail at her, oh my, that was it. And with the frustration of being prevented from seeking out that social play, she would become vocal, often appearing quite aggressive! They were very concerned as this behavior developed around two years of age, until they realized the play signals. And when other (well-behaved) dogs did come close, even on leash, she was simply all about play.

We would have had to raise the costs (aversive responses) to extreme levels to overcome the benefit (in her mind) of socializing. So instead, we trained, and trained, and trained, an alternative behavior, lying down. If she saw a dog, at the right range, before over-stimulation, then she learned to lie down, get a lot of great rewards, and this calmed her down and allowed better management of the situation.

Again, he viewed this all as another example of the relative costs and benefits of behaviors, in the perception of the dog.

CHAPTER 7

You can't always get what you want: The costs and benefits of being social

TIME IS ON MY SIDE

The voice broke the stillness like a rock shattering the glassy surface of an iced-over lake. Ziva's hackles raised as the unfamiliar utterances intruded into their blind. She bumped her nose against Alec's arm with increasing urgency, watching his face closely as he woke. Alec placed his hand on her head as he stood groggily from his chair, fixing his gaze in the direction of the voice. The sun directly overhead filtered light through the foliage with splashes of green and gold. Ziva growled quietly as the unfamiliar voice was joined by another, and then two bipedal figures emerged from a copse of trees, their arms gesticulating animatedly as they spoke. The voices reached a crescendo as the two men argued; Ziva's growl grew louder until Alec quieted her with a gentle tap to her shoulder and a pointed, "Shhh." The two men continued on, their voices gradually fading as they crossed the meadow below. There was movement in the thicket flanking the meadow and a startled bird flew into the midday sky. The men paused and contemplated it before they continued on. Ziva's mouth opened noiselessly and she eyed the bird eagerly, moving forward with her tail raised, but then she regarded Alec as the two strangers finally disappeared into the trees once again. Alec glanced down at her, patting her shoulder, and she reclined onto the earth once again.

    This seemingly mundane interaction between a man and his dog actually reveals a complex interplay of social behavior and genetics that's tens of thousands of years in the making. While we may take for granted our dogs' innate abilities to read our gestures, note changes in the tones of our voice and our eye direction, and comprehend both words and meaningful non-word sounds such as "shhh," this actually exemplifies an intricate interspecies social relationship that is unparalleled in the animal kingdom. Humans, like domesticated dogs, are highly social animals who have learned to communicate with one another with very subtle cues, resulting in the most successful interspecies relationship on record. But there are both costs and benefits to being social – or not – and maintaining these relationships (or not). And individual temperament and personality both play important roles in these relationships.

    Not all animal species are inherently social; the degree of sociality varies widely. Some animals prefer to spend their time among the company of other animals, both inter- and intra-specifically. Some animals, however, prefer to be solitary for the majority of their time, excepting time spent mating and raising dependent young. Solitary vertebrate species include, but are not limited to, bears, badgers, platypuses, sea turtles, Tasmanian devils, moose, koalas, and many species of feline, including mountain lions, tigers, leopards, bobcats, jaguars, and ocelots. Solitary animals tend to be territorial and will fight, often to the death, to defend their resources, including food and opportunities for mating.

There are costs to being more social, including increased competition for resources such as food, shelter, and mates. Animals such as lions, meerkats, chimpanzees, and wolves often work together to acquire food, but then they have to share these resources. If three lionesses on the plains of the Serengeti worked together to take down a gazelle, they don't get to consume it first; the male lion in their pride gets access to the meat, followed by the females and their dependent young. Male lions rarely help with hunting, which involves coalitions of female pride members, but males do provide protection from other male lions who would likely kill the pride's cubs during a takeover. Social living also incurs an increased risk of infection by disease and parasites. Close social living means easier disease transmission, while solitary living means decreased opportunities for an individual to catch a disease or spread a parasite. In 1348, 14-year-old Joan Plantagenet departed Portsmouth, England for Castile, accompanied by an entourage of 150. Young Joan of England was traveling by way of Bordeaux to meet and marry her fiancée, Pedro, but the two never met; Joan died from the Black Plague before she reached the border of her betrothed's country. In all, the Black Plague of the 14th century would sweep through the once densely-populated portions of the European continent, causing the deaths of 50 million people – roughly 60% of Europe's entire population. Diseases like the Black Plague, yellow fever, smallpox, typhoid, and influenza are highly contagious and don't discriminate between young and old, rich or poor (Fig. 30). They only need the right conditions – including high population density – to flourish. The plagues of old have long been blamed on rodents that were carrying fleas and ticks, but new evidence points to humans as the possible carriers – and unwitting forms of transportation – for these virulent vectors. While some people have genetic immunity from some diseases, such as malaria, and vaccines have been created to eradicate many of the worst diseases this planet has seen, pandemics have extended into modern day. Diseases like yellow fever are spread via mosquito bite; higher population densities means more opportunities for mosquitoes to transmit the disease from person to person. Throughout the 17th, 18th, and 19th centuries, yellow fever was one of the most infectious diseases in the world. While it originated on the African continent, it spread to Asia and to the New World with the slave trade. In 1878, yellow fever struck Memphis with a vengeance, resulting in 17,000 cases and more than 5,000 deaths. Today, yellow fever is still one of the most deadly diseases in India. A high population density, paired with more than 1 billion people residing in India contribute to its rapid spread. Contagious diseases aren't just the plight of humans, though. Among dogs, canine parvovirus is a highly virulent disease that can survive in the environment for months; for canines with naïve immune systems, the disease can spread rapidly, often with deadly results. Parvovirus is most common in urban areas with high canine population densities, and given its highly resistant, contagious nature, is easily spread outside the main area.

FIG. 30

Black plague mask.

Living in a social group also means increased risks incurred pertaining to infants, including increased risk of higher infant mortality and the increased risk of caring for non-offspring, also known as "alloparenting." In a larger social group, infants run the risk of disease transmission and injuries from other members of the group far more than they would if they only lived with their mothers. The Comparative Breeding Hypothesis states that alloparenting was integral for reproductive success for humans during the Pleistocene,[117] but caring for the offspring of another group member would be calorically and temporally expensive, even if the offspring had a degree of relatedness to the caregiver.

---

[117] Hrdy, S. (2005). In C. S. Carter & L. Ahnert (Eds.), *Attachment and bonding: A new synthesis. Dahlem Workshop No. 92*. MIT Press: Cambridge.

The benefits of being highly social are also numerous. With increased group size, there is a decrease in predation pressure due to detection, repulsion, and swamping. While the three lionesses of the Serengeti had to share their kill with the male lion, they also benefited from his protection against the threat of other males, who would likely kill any young offspring in the pride during a takeover. There is also an increase in the defense of resources and the benefit of increased care of offspring via alloparenting. Alloparental care is demonstrated in 120 mammal and 150 avian species, including 20% of the primate genera (including humans),[118] lions, meerkats, jackals, and wolves. Alloparental care affords the biological parents of the offspring the opportunity to hunt and forage, which is particularly important if the parents are also the most successful food providers in the social group. With many social structures, larger social groups also provide the added benefit of increased availability of mates, and thus more opportunities for genetic diversity of offspring.

The costs and benefits of being social exist on both the individual level and on the species level. On a species level, both humans and canines prefer to spend a large portion of their time socializing with their peers. Humans interact with one another during their professional lives, social clubs, sports, hobbies, and school. Humans have spoken, written, and signed languages to communicate with one another; we have pictographs, hieroglyphs, memes, gifs, emoticons, and slang to clarify the context of their messages. And technology, including email, phone, social networks, and Skype, have enabled humans (and sometimes even their dogs!) to maintain their social relationships, regardless of the distance. But along the way, dogs had to weigh the costs and benefits of this social relationship with humans – and whether it was "worth it" to them to become domesticated. On an individual level, each animal has a different temperament and chooses just how social they want to be.

These costs and benefits can be partially explained in humans' and dogs' shared architecture for sociality. For both species, evolutionary processes favored the development of complex social behaviors. This is also true for humans' closest living relatives, the chimpanzees and bonobos, from which we diverged approximately 6 million years ago. *Homo sapiens*, along with chimpanzees and bonobos, belong to the Homini genus. According to American biologist, ecologist, and anthropologist Dr. Jared Diamond, genetic evidence provides support that humans should be classified as the third chimpanzee species, along with common chimpanzees and bonobos. Diamond's groundbreaking 1991 book, *The Third Chimpanzee*, challenged the current taxonomy of the time, noting that the chimpanzee-human genetic differences are actually smaller than some animals' within-species genetic differences. The social complexity exhibited by chimpanzees and bonobos (including tool use and creation among chimpanzees and widespread cultural differences between geographically separated social groups among both species) only adds to this hypothesis, but chimpanzees and bonobos are not as highly encephalized as humans. Humans are highly encephalized, meaning that their brain size is larger than would be expected, given

---

[118] Riedman, M. L. (1982). The evolution of alloparental care and adoption in mammals and birds. *Quarterly Review of Biology*, 57(4), 405–435.

their relative body sizes. The human neocortex is significantly larger than would be expected, given the neocortex to body size ratio of many mammals. This is particularly important because within the neocortex resides the areas of the brain devoted to language, social cognition, emotional regulation, conscious thought, empathy, and theory of mind – the ability to comprehend the perspective, intentions, and feelings of other beings. Chimpanzees have an Encephalization Quotient (EQ) of 2.2 to 2.5, below that of the Orca (2.57-3.3), bottlenose dolphin (4.14), and human (7.4-7.8) (Fig. 31).

Of all known living beings, humans are uniquely hardwired for social living – and our closest companions, having coevolved with us for millennia, have evolved convergently and share many of these traits with us. While dogs and humans belong to taxa that are separated by 100 million years of evolution, the brains of humans and dogs have functionally analogous regions, including voice-sensitive cortical areas. Ziva might not have understood all of the words that the two strangers were exchanging, but she was processing it similarly to Alec, and she, like Alec, could deduce that the conversation was emotionally intense. It isn't just our neural architecture that we share in common with our canine companions, though; while there's millions of years of evolutionary separation between us, we have other genetic traits in common, as well. Researchers from the University of Chicago found that several groups of genes in humans and in dogs have been evolving in parallel for millennia. This likely began to occur at the onset of domestication, during which time dogs who were more social and whose personalities were cohesive with their human counterparts were preferentially selected for in comparison to their less social peers.

Examples of EQ values

Species	EQ
Hummingbird	9
Human	7.4
Dolphin	5.6
Killer whale	2.9
Chimpanzee	2.5
Rhesus monkey	2.1
Elephant	1.9
Whale	1.8
Dog	1.2
Cat	1
Horse	0.9
Sheep	0.8
Mouse	0.5
Rabbit	0.4

EQ is an estimate of species' intelligence based on brain/body size. It is the ratio of brain weight of a species to brain weight of an average species of same body weight. An EQ > 1 means that the species has a larger brain than expected from simple body weight; an EQ < 1 means a smaller brain than expected for body weight.
Based on Jerison, 1973.

FIG. 31

Encephalization quotient (EQ) of various animals.

LET ME BE YOUR BEAST OF BURDEN

Why would an animal choose to go from a life of relative freedom and choice to a life of domestication? What were the conditions that created the perfect environment for the ancestors of certain species, including modern horses, cattle, and dogs, to become beasts of burden and constant companions? There was likely a perfect storm of conditions, wherein those animals who became domesticated did so because the benefits outweighed the costs. Domestication, which is often correlated with human population density increases and crowded living conditions, likely occurred with certain animal species as the ancestors of modern man began to stay within one geographic area. These novel environments, paired with our innate sociality, might have been the selective pressures that drove the rewiring of both humans and dogs. Living in crowded conditions with humans may have conferred an advantage on those dogs who exhibited less aggression than their less gregarious peers; these dogs would have been preferentially retained and bred with other non-aggressive dogs, eventually leading to a lineage of friendlier, more submissive canines, just as Belyaev demonstrated with his foxes (Fig. 32).

Aesthetically pleasing, dark haired and eyed, and with a shared intensity for their work, Dmitri Belyaev and his partner, Nina Sorokina, were the Cold War's version of Skully and Mulder. Ensconced in the far reaches of Siberia, Belyaev and Sorokina set out to demonstrate how domestication may have occurred with dogs, tens of thousands of years before, but not without the kind of masterful maneuvering that would've impressed Indiana Jones himself. In his book *How to Tame a Fox (and Build a Dog)*, Dr. Lee Dugatkin provided the first complete overview of this study, noting that Belyaev believed that it would elucidate the mystery behind domestication, asserting that humans were merely "tamed apes" (and tamed chimpanzees, Jared Diamond would further argue.) Despite the potential importance of this

FIG. 32

A wild fox and a tamed fox. *From Kayfedewa at English Wikipedia.*

study, genetic research in 1950s Cold War Russia wasn't easy to pioneer or patronize. Belyaev's experiment was equal parts scientific exploration of domestication and Cold War political thriller: to perform his longitudinal study on modified descent with foxes (with the hope that successive generations would be domesticated), he had to disguise the study. Rather than saying that he was investigating domestication with another species of canid, he told the authorities that he was exploring how to increase the number of fox pups born and the amount of fur that they produced – a goal that would be financially lucrative for Russia. Dugatkin argued that this was a front that not even a dictator could argue against. Beginning with a founding population of a few dozen silver foxes from Russian fox farms, Belyaev and Sorokina began to breed the calmest, least aggressive foxes with one another. Initially, the foxes remained stand-offish with humans, but over the generations, they were eventually rewarded with a series of increasingly more docile foxes, most notably beginning with Pushinka in 1974, who would become just as loving and loyal as any domesticated dog.[119] Succeeding generations of foxes not only tolerated handling more than prior generations, but they also showed affection with humans, sought out human contact, began to whimper and lick people like dogs did, and they began to wag their tails, too – a behavior that's not seen in wild foxes. This selective breeding shaped the future of domesticated dogs who, in turn, influenced the future direction of man.

While our neural networks were separated by tens of millions of years, over time, humans and dogs began to exhibit similar brain wiring and behavior. And while selection in the same gene with two different species (convergent evolution) is an evolutionary rarity, it's not unheard of, especially in the case of species that have shared or similar environments or two species that have co-evolved for so many thousands of years. Examples of convergent evolution include sharks and dolphins (cetaceans), who share the same marine environment and many of the same morphological features, but are not closely related. Cetaceans' closest living relatives are actually the hippopotamuses; their ancestors diverged from their lineage 40 million years ago.

Humans and canines share many genetic commonalities. The gene sequences for the transmitters of serotonin, cholesterol processing, and cancer have been selected for in both species, and as a result, they share multiple diseases in common, including some cancers, obesity, and epilepsy. With similar neural architecture, humans and domesticated dogs also suffer from many of the same psychological issues, including depression, anxiety, obsessive-compulsive disorder, and post-traumatic stress disorder. This is possibly due to the fact that genes can often have multiple effects, which is referred to as "pleiotropy."

YOU CAN'T ALWAYS GET WHAT YOU WANT

Gregor Mendel first observed the phenomenon of pleiotropy (pleio is Greek for "more" and tropic means "way") during his pea plant inheritance studies. Mendel found that plants with colored seed coats would always have colored leaf axils and

---

[119] Dugatkin, L. (2017). *How to tame a fox (and build a dog)*. Chicago: University of Chicago Press.

colored flowers. Pea plants that had colorless seed coats always had unpigmented axils and white flowers. There was an association between seed coat color and axil and flower coloration. Pleiotropy is the phenomenon where one gene can influence two or more phenotypic traits. Think of this as the evolutionary version of car buying: sure, you can have the vehicle with arctic pearl white paint, but only if you also have the dove-colored leather seats and the platinum woodgrain package. In the case of the seed coat color genes, the gene controlled not only seed coat color, but axil and flower pigmentation, as well. Among animals, examples include blue eyes and deafness in domesticated cats (40% of cats who have blue eyes are also deaf); and the foxes in Belyaev's longitudinal study, wherein subsequent generations were more docile than their ancestors, but they also had a slough of new phenotypic traits, including distinctly spotted coat coloration and floppy ears. Among humans, examples of pleiotropy include sickle cell anemia, wherein the individual is not susceptible to the effects of malaria, but their body also doesn't have enough red blood cells to adequately oxygenate it. Thus, some of the effects of a pleiotropic gene mutation can be beneficial, while others can be deleterious; genes are selected for when the selective benefit outweighs the cost.

For humans, the benefits of a social brain and social living a highly complex social life are numerous: larger social groups provide more protection from threats, both external and internal. Maintaining social connections are vital for most humans; there's a reason why solitary confinement is considered to be the most austere of our punishments within the judicial system. When humans share social interactions, whether it's with other humans or with nonhuman animals, we release oxytocin, which is colloquially referred to as the "love hormone." Oxytocin has been associated with pair bonding, increased fidelity, and mother-child bonding. It's a chemical "reward" that makes social bonding feel good. The importance of connecting socially is reflected in our abilities to empathize and sympathize with other beings and with how we speak about our emotional relationships; many languages have expressions like "hurt feelings" that compare the pain of such social rejection to the pain of physical injury. But "hurt feelings" goes far beyond a playground metaphor; social pain is associated with the anterior cingulate cortex, which is also involved in the emotional component of physical pain. Social pain, like social pleasure, has a physiological component, as well.

While humans are highly social beings with vast cognitive capacities, there is an initial cost to this. Humans aren't precocious when they're born; infants who develop over the course of an average gestation have encephalized brains that take up more than 100% of the mother's birth canal during birth. Even then, their brains are still "underdeveloped," in comparison to other primates. Humans have a chart-topping Encephalization Quotient (EQ) range of 7.4 to 7.8. In comparison, their closest living relatives, the chimpanzees, have an EQ of 2.2 to 2.5. If the brain was allowed to grow more in utero, resulting in a more precocious baby, then babies wouldn't be able to pass through the pelvic canal. Humans have a prolonged infancy and adolescence as their brain continues to mature, during which time they are very dependent upon their parents. Prey species, such as horses, typically stand within an hour or less of birth. Horses have an EQ of .9. Comparatively, human infants usually stand unaided between 8 and 12 months of age; dogs, who are born deaf and blind, typically first attempt to stand and walk at three weeks of age. (While dogs have an EQ of 1.2, it's interesting to note that wolves' brains are typically 30% larger than the brains of their domestic cousins.)

Human parents are responsible for their offspring beyond the timespan of any other primate. In comparison, our closest genetic relatives, the chimpanzees and bonobos, with whom we share 98.6% of our DNA in common, start to become independent from their mothers around age 6, reaching full adulthood and autonomy by the age of 15. Gorilla infants spend only 50% of their time with their mothers by the age of 30 months, with full autonomy around the age of 6. Orangutan infants are typically fully independent by age 10. Exceptions of prolonged over-reliance, such as Jane Goodall's chimpanzees Flo and Flint, are noteworthy outliers. Venerable Gombe matriarch Flo had given birth to at least three offspring: Faben, Figan, and Fifi, before she had her son, Flint. While most chimpanzees begin to become independent around the age of six, Flint resisted well into his eighth year. Flint was still not weaned from Flo even after she gave birth to his sister, Flame, when he was eight years old, and Flint insisted upon still sleeping next to his mother, riding on her back, and spending all of his time with Flo. After Flame died at six months of age, Flint became increasingly reliant upon his mother, and when Flo died in 1972, Flint was unable to survive without her support. While chimpanzees are said to be fully adult by age 15 and humans are "legal" adults in the US at the age of 21, the human brain is not fully developed until the age of 25. Adults think and process information using their prefrontal cortex, while teenagers do so with their amygdalas, which are their brain's emotional control centers.

Humans don't just parent their biological offspring for a relatively long period of time; they do so with their pets, as well. Domesticated dogs also have an artificially elongated adolescence, as do cats, to a certain extent. Humans often see themselves as "pet parents," and subsequently infantilize their pets, treating them as infants throughout their lifespans. Their dogs rely upon them for all of their needs, including food, water, shelter, and social enrichment.

BUT IF YOU TRY SOMETIMES, YOU GET WHAT YOU NEED

Many of the members of any given social group are blood relatives, but the success of a social group requires that all members live cooperatively, to a certain extent. A group cannot be cohesive if there is constant strife within it. So what mechanisms can explain the cost-benefit of being social among non-related individuals? Kin selection can explain the evolution of altruism and altruistic behaviors among closely related animals. The Coefficient of Relatedness, r, can be calculated as follows: Identical twins have an r of 1, parent-offspring have an r of .5, as do full siblings, half siblings have an r of .25, as do nieces and nephews and grandparents, and cousins have an r of .125. Under Hamilton's Rule, kin selection is favored when $r \times B > C$, with B referring to benefits, as measured by offspring equivalents, and C referring to costs, as measured by offspring equivalents. For example, a lioness who nursed her full sister's starving cub would gain inclusive fitness by this act, as the benefit, B, to her sister that the cub would survive would outweigh the cost, C, of the sister's caloric expenditure and slight reduction in what her own cub(s) ate. Thus, seemingly altruistic behaviors, when exhibited by closely related individuals, can be explained in evolutionary terms.

Cooperation between non-relatives, however, is typically explained through other frameworks of natural selection. Truly "altruistic" behavior is considered to be a rarity, but reciprocal altruism can be understood with the premise that "if you scratch my back, I'll scratch yours." With reciprocal altruism, there's also the added element of temporality – the initial provider of a seemingly altruistic act pays an immediate cost for the recipient's immediate benefit, but then banks on their own future benefit. This leaves a loophole for selfish individuals: the recipient of the initial altruistic act could potentially fail to return the favor, having gained that first initial benefit. They would have a net gain at no cost – at least for the moment. The provider, of course, would be less likely to help them in the future, having learned that this individual does not reciprocate. But if the two individuals worked together for their mutual benefit, then this would be cooperation. Even with the possibility that there will be mutual benefit, there's still the chance that cooperation will not evolve.

Animals are constantly considering the relative costs and benefits of their actions, and in the practical, clinical world, we can benefit from understanding the dog's perception of these costs and benefits. For instance, you can frequently gain insight into social behavior problems at off-leash dog parks by carefully considering these trade-offs in my subject. A common issue was exhibited by a poodle named Blue who seemed to love arriving at the park, but he became involved in nasty dogfights once he arrived. The owner was both perplexed and convinced by Blue's eager anticipation that going to the park was a good thing for Blue, but then, how was he supposed to deal with the fights?

An assessment revealed that Blue was a highly social animal, in terms of basic personality, but that he became quickly overwhelmed by the intensity of social interaction. He became anxious as the situation grew out of control (at least in his perception), and his evaluation of the relative costs and benefits quickly changed. Yet, in instances in which the park was quiet, he was able to establish tolerable social interactions, which he valued highly. So, the park visit remained relatively high-value for him, but intense, over-(his)-threshold interactions were perceived as a serious cost, and he switched to defensive behavior to try to regain control. These cost-benefit calculations are occurring continuously in our dogs (as well as ourselves!)

The correction was to recommend stronger owner management of the situation: Use of the parks at less-crowded times, visiting smaller parks, and ensuring more structured social interactions like organized play groups or dog daycare facilities, but I also taught the owner to perceive the surroundings like his dog, and to watch his dog's body language to determine how his dog was calculating those relative costs and benefits.

Understanding how an individual perceives their surroundings is key to understanding how they calculate relative costs and benefits. A lot of weight (perhaps too much) has been placed on personality tests that seek to categorize us based upon a number of factors that are intended to reveal how we perceive, process, and socially interact with the world around us. Are we extroverted or introverted? Sensing or intuitive? Feeling or thinking? Judging or perceiving? The Myers-Briggs Type Indicator (MBTI) is a self-reporting questionnaire that categorizes how individuals perceive the world and make decisions in one of 16 different categories (Fig. 33). The MBTI is based on Carl Jung's concept that humans experience the world through thinking, feeling, sensation, and intuition and it provides participants with a four-letter label, like ENFJ, ISTP, or ENTP, that summarizes how they perceive the world. Not everyone agrees with their designations, though. Many are reticent to self-label as "introverted" – in a highly social society, extroversion often receives more esteem than introversion. In the book *Quiet: The Power of Introverts in a World That Can't Stop Talking*, author Susan Cain expounds upon the recent rise of the extrovert and the undervaluing of the introvert, even though introverts account for approximately one-third of the population.[120]

Similarly, many of the more popular dog breeds in the United States, such as Labrador retrievers, golden retrievers, and poodles, are described as "friendly,

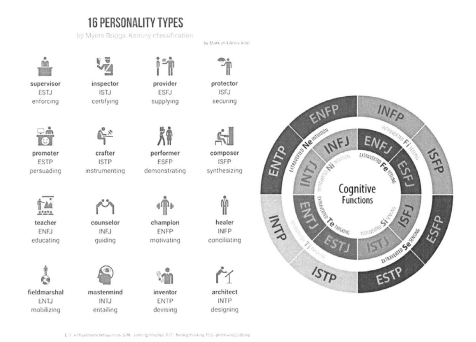

FIG. 33

The Myers-Briggs personality types.

---

[120] Cain, S. (2012). *Quiet: The power of introverts in a world that can't stop talking*. Random House.

gregarious, and outgoing." Shelters with adoptable animals often focus on these traits to help them find a home. These are highly desirable traits for amicable social living, and while generally accurate for these breeds on the whole, these descriptions overlook individual variability. Humans and dogs are both very social beings, but individually, there is a continuum of extroversion, friendliness, and social skill. We've all been at a social event with a person who didn't understand social graces – someone who talked while standing too closely, didn't understand disengagement, or fired out a series of increasingly uninteresting non-sequiturs. Similarly, not all dogs share the same level of social skill; those who have been raised without the socialization of humans and other dogs, thus missing out on critical social periods, may struggle with socialization.

Temperament adds another dimension to the valuation of perceived costs and benefits in an animal's behavior. Just when you begin to grasp that there are costs to every action, and perceived costs to those actions, and that the relative balance of those two sides determines behavior to a great degree, the genetic component must be added, as well. And the strongest genetic influence, by far, is temperament, or personality if you prefer: those independent dimensions of predictability in behavioral response that arise from both genetics and past experience. Is your dog, or spouse, or child, shy or bold? Aggressive or retiring? Introverted or extroverted? Cautious or carefree? Fearful or without fear? Or of course, somewhere along the continuum of each independent dimension.

Why does personality matter so much? "Personality" generally refers to an individual's suite of characteristics and the individual differences that one has in how they behave, feel, think, and process. While science was initially slow to accept that animals other than humans even had "personalities," it's now widely accepted that nonhuman animals, like humans, have a wide variety of personalities and temperaments. As of this writing, more than 100 animal species have been said to have distinct personalities; it's only a matter of time before that list grows. University of Washington PhD student Carly Loyer, whose degree focuses on applied behavior, studies personality in dogs, but it's still a rather amorphous research field.

"I don't think we know nearly enough about personality in dogs!" she said. "We know enough about personality in humans and other animals to say that canine personality could affect dogs' responses to various training techniques, their aptitudes, and their social behavior, including aggression. I have yet to see personality research in canines that I feel truly captures the dimensions of personality that I want to understand, though there are many interesting research teasers that invite follow up."

"Current research is examining the links between behavior, breeds, and genetic markers has recently caught my attention. We can assign mean, general, fundamental personality characteristics to breeds or breed groups. I think we can assign some general behavioral traits to dog breeds, particularly working lines or breeds that are still selected for their behavior," Carly explained. "The Russian silver fox experiments have demonstrated pretty clearly that breeding for behavior can create drastically different behavioral phenotypes or temperaments. That said, as breeding emphasis veers more towards looks and less for behavior, it's much more difficult to say what behavioral traits come along for the ride (and the silver fox experiments have also shown that there are links between looks and behavior). If curly tails and floppy ears are linked to human sociability, what behaviors might be linked to snub noses and oversized foreheads?"

Temperament and personality are often terms that are used interchangeably in animal behavior research circles to avoid anthropomorphism. To avoid this, personality researchers use behavioral criteria to measure "personality," such as whether an individual consistently shows a particular trait such as boldness or timidity over space and time. If an animal has the same response to a situation each time (say, running away from a potential predator or approaching the potential predator) they can qualitatively be classified as "timid" or "bold" (and perhaps fool-hardy, as well, if they're a prey species that's always approaching a dangerous predator). But perhaps this "bold" strategy was adapted to a specific environment – one where advancing, rather than retreating, conferred a selective advantage to those individuals that did survive a potential altercation with a predator.

Dogs, like many other animal species, have certain personality "types," including confident, laid back, independent, timid, and adaptable. Personality traits are heritable, as well. If someone has ever told you, "You act just like your mother," or "Your father reacts the same way to these situations!" it might just be in your DNA. That's not to say that everything is predetermined, though; your genetics provide a predisposition for certain traits, but not necessarily others. Why certain similar traits arise in an individual's offspring and not others is a mystery. In 2004, undefeated racehorse Smarty Jones captured hearts nationwide as he tried to win the elusive Triple Crown. While he was caught by a charging Birdstone and fell a length short in the Belmont Stakes, retiring shortly afterward, he remained a favorite – and he was an exciting sire prospect. While a lot of Smarty's offspring didn't seem to have that same drive to win, what they did seem to inherit from the handsome red horse was his intelligence and his temperament. Heritability is potential, but not a guarantee. Smart Dixie Jones, one of Smarty's sons, won only one of his 23 starts and finished last in most of his races. He was fast, and his sole triumph occurred with ease, but he didn't seem to *want* to run. What he did inherit, though, was his sire's other personality traits. He wasn't spooky, but approached everything with a "been there, done that," attitude. He was kind, intelligent, and constantly "busy." Perhaps these traits were more dominant than those that made his sire a successful racehorse; and perhaps the genes that helped Smarty Jones win all but one race will be expressed more in other crosses or in later generations. Perhaps like Secretariat before him, Smarty Jones will prove to be a popular influence in the breed several generations later.

Dogs, too, don't always inherit what's "expected" of them. You can get water-resistant Labradors, like Jack (who might get his water aversion from his Border Collie heritage, or perhaps he just missed the memo on heritability of predilections and personalities), hunting dogs that bark too much, and dogs from a "therapy dog" lineage who are too hyper to perform their duties. But by and large, certain breeds do seem to have certain traits and personalities that distinguish them from other breeds.

GOOD TIMES, BAD TIMES

Is the bowl half empty or half full? Judgment bias refers to how an individual perceives reality, based upon their experiences and perceptions. Examining an individual's judgment bias can determine whether they're optimistic or pessimistic.

"We need to study judgement bias in dogs for two reasons: first, it's something that we can impact in some pretty straightforward ways, and second, it's something that can probably impact dog behavior," explained Carly Loyer. "Judgement bias represents what may end up being a very useful feedback loop. If inciting a more optimistic view of a social situation can improve a dog's behavior in that situation, the improved interaction that follows may affect the dog's judgement bias in the future, and so on… but we don't know about all of those "ifs" yet, and we should! On the other hand, if a dog's misbehavior stems from an overly optimistic judgement of social situations (overly optimistic about their ability to deter an opponent, or overly optimistic about another dog's desire to interact with them), we may need to change our training tactics to help them assess social interactions more evenly."

The best canine-human relationships don't have dissimilar personalities or temperaments; for example, an avid runner might be better paired with a greyhound than they would be a bloodhound, and someone who was looking for an agility work partner would be better suited with a border collie or an Australian Shepherd than a toy dog.

Temperament can play a very important role in an applied setting, as well. The role of temperament, and the evaluation of temperament in case dogs, is primarily to inform us as to the tendencies for a subject. Was this dog a Bold, Outgoing, Extraverted pup to start with, likely to stick his nose into places it might not be wanted? Or perhaps she was Introverted, Anxious or Nervous, or Shy to start with? These built-in personalities need to be factored into the behaviors JCH is seeing as part of the case evaluation, or into what he is hearing about what happened during an incident. The temperament also gives him information about directions to take in treatment recommendations: how will a dog accept corrective actions, how easily will a dog accept social interactions, even if rewarded?

Finally, there are cases where he has to have a chat with an owner about lifestyle, theirs and their dog's. We know breed temperament and we can confirm individual temperament, and discover that a low-speed owner has a high-speed dog, and even a low-speed (not built for exercise, mentally or physically) dog with a marathon-training owner. How about a fundamentally-not-very-social dog trapped in a downtown condo full of dogs? This is happening more and more often: we can try some good rewarding of social behavior, but we often cannot fundamentally change an asocial temperament to a social one. And all too often, I see dogs being nearly dragged around a jog… not ALL dogs need more exercise, just a LOT of them.

While the science of evaluating dog personalities is still in its infancy, human personalities have been measured in various ways, including the MBTI, which was further classified in Dr. David Keirsey's "four temperaments." The four temperaments divides the MBTI's 16 constructs into four groups: the Guardian, the Artisan, the

Idealist, and the Rational.[121] One study has also categorized the personality of children across nine basic criteria: sensitivity, activity level, intensity of reaction, rhythmicity, adaptability, mood, approach/withdrawal, persistence, and distractibility.[122] Similar attempts have been made to classify the temperament of dogs, including a 2003 American Veterinary Medical Association (AVMA) study that measured temperament traits, including aggression, fear, chasing, trainability, and attachment.[123]

JCH says that his evaluation of a dog's temperament, based on owner report, his observations, and often, his active assessment is probably one of the single biggest factors in my evaluation of a behavior issue or problem. How did this dog react to the stimuli at the time of an incident? How will it react now? How will it react to training or management put into place to correct a problem? When it comes down to it, he is asking: How does this dog weigh its costs and benefits? Remember, it is the *perceived* costs and benefits that produce behaviors, and it is temperament that alters those perceptions.

As I (JCH) go into a case, whether it is Calvin the stranger-aggressive Cattle Dog, or Cheeto the house-soiling Whippet, it's the dog's temperament that I want to quantify. Now, of course, I need to understand the breed-typical behaviors, the early history and potential-traumas of the dog, the household structure (physical and social), the temporal and spatial patterns of the incidents, and changes in any of these factors. But all of these factors and their effects on the patient are all funneled through the filter of the individual temperament. Calvin was not very bright, but very stubborn; he was an outgoing, aggressive, bold, forward-leaning, high attachment sort of temperament type (a type is a common cluster of temperament factors): when he finally realized that the (generally rare) arrival of strangers took owner attention away from him, he reacted "aggressively" and discovered that it worked… strangers backed away and the owners gave him attention. He had altered the costs and benefits of stranger visits to his favor…he quickly latched onto this learned behavior. The treatment: the owners needed to take back control of those costs and benefits…make the behavior more costly (punishment) or better in my mind, and less dangerous to all, make an alternative behavior more beneficial: reward good behavior. Make sure Calvin got attention even though strangers are present. It worked like a charm! The owners thought that they would be rewarding bad behavior but instead, Calvin simply needed reassurance.

Cheeto was the opposite in many ways: fearful of his own shadow. Once we determined that the house-soiling was not due to any health issues, we began to assess his temperament more completely. Clearly, he was fearful, not social at all, but he was also very bright. He was home alone all day, but exhibited no barking or house damage or soiling early in the day, right after owner departure but rather, he would

---

[121] https://keirsey.com/4temps/overview_temperaments.asp.

[122] Kristal, J. (2005). *The temperament perspective: Working with children's behavioral styles*. New York: Paul H. Brookes Publishing Company.

[123] Hsu, Y., & Serpell, J. A. (2003). Development and validation of a questionnaire for measuring behavior and temperament traits in pet dogs. *Journal of the American Veterinary Medical Association*, 223(9), 1293–1300. https://doi.org/10.2460/javma.2003.223.1293.

be destructive later in the day, or even after the owners arrived home, and right in front of them. He failed miserably at dog daycare and with dog walkers, being mostly afraid of them all. The owners thought that he was afraid of them (and I probed diligently for any reasons why Cheeto might feel that way, to absolutely no avail). Noting from owner reports that Cheeto was a high-functioning problem-solver, I took a guess and suggested that he was exhibiting attention-seeking behavior and boredom. We put Cheeto on a dramatic work-to-eat food puzzle program, which occurred all during the day and involved automation, hidden food, and even music and video that went on and off throughout the course of the day. Problem solved. Cheeto was basically bored out of his gourd during the day, and he was emotionally wound up by the time the owners came home. As a result, he was nervous about his daily walk, and yet he was seeking attention (stimulation) from his owners. All of this put him over-threshold and he would act out and house-soil.

These examples illustrate the importance of temperament, and the dog's perception of their world and its costs and benefits, to solving real-world behavior issues.

GIMME SHELTER

Of all man's innovations, the most influential just might be domestication. There is fossil evidence that indicates the beginnings of a mutual relationship between humans and nonhuman animals as far back as half a million years ago. The selective breeding of a number of species, including horses, sheep, cattle, and dogs, has resulted in dramatic changes in our technology, agriculture, and social lives. Domestication is correlated with higher population density, sedentism, reflecting significant changes in lifestyle, technology, and worldview. Land was now viewed as a commodity; one that needed to have resources extracted and that needed to have an animal workforce to help till fields, uproot trees and pull downed ones, mine, and provide protection from outside threats.

Domestication is the permanent genetic modification of a species. Domesticated animals are genetically and behaviorally different than their free-living cousins. The ancestors of modern humans selected certain species for domestication based upon a number of factors, including their social structure, their interactions with humans, and their flexibility and reactions to novel stimuli and environments. While humans were choosing these animals, though, the animals had to choose domestication, as well; not all animals have the temperament to be domesticated, and domestication has operated on comparatively few previously wild species. According to Jared Diamond,[124] there are six criteria that an animal must meet to be successfully domesticated. First, the animals must not be particular about what they eat, but be more flexible with their diets. Dogs belong to Carnivora, but they are able to eat a wide variety of foods, from scavenging to even eating a vegan diet. Second, animals must reach maturity quickly, relative to their human counterparts. Elephants are

---

[124] Diamond, J. (1997). *Guns, germs, and steel.* New York: W.W. Norton & Co.

large, long-lived, highly capable beings, but they also have a very long time to reach maturity, relative to other animals, such as horses. Thus, horses and cattle became domesticated on a wide scale, while elephants were tamed with relative rarity. Third, an animal must be willing to breed in captivity, to create future generations of domesticated individuals. Fourth, the animal must have the right temperament. While the horse has been domesticated for thousands of years, his tempestuous cousin, the zebra, has only been tamed (and not domesticated) on rare occasions. Fifth, domesticated animals can't be prone to panicking and fleeing when faced with a potential threat. Horses and cattle will typically exhibit a startle response, but many wild ungulates have an inexorable urge to flee when startled, making them inappropriate for domestication attempts. Sixth, domesticated animals must conform to a social hierarchy that has a strong leader or parent figure. Horses are prey animals who want a rider who will both literally and figuratively take the reins. Dogs are highly social beings who are looking to the humans in their household as the parental figure. The only exception to this last criteria is the domesticated cat, whom many would argue is still only "partially domesticated."

When Ziva's ancestors began to spend more of their time with their bipedal friends, they were influenced by the costs and benefits of this lifestyle change. For a few select species, the benefits of domestication, including shelter, food, protection, companionship, stability, and consistency, outweighed the costs of decreased choice and living under another species' set of rules. Humans, too, incurred both costs and benefits when they domesticated animals: in exchange for added labor, companionship, and protection, they were now responsible for feeding, sheltering, and providing care for additional members of the "family." Over time, these once-independent species became wholly dependent upon humans, exhibiting both behavioral and morphological changes that reflected their new alliance with humans, but that would be maladaptive for living in the wild. For the domesticated dog, these changes included decreased head size in relation to wolves, smaller jaws and teeth, reduced aggression, bite inhibition, prolonged adolescent periods, a longer primary socialization period, and the retention of neotenous characteristics, even after adulthood. While these traits would be maladaptive for living in the wild, they were highly adaptive for living with another social species – and particularly attractive to humans.

CHAPTER 8

How I behave depends upon how you behave…maybe: Game theory and our canine companions

BEAUTIFUL MINDS

In 1928, a 25-year-old Hungarian mathematics student ascended the pulpit to give his first public presentation. Brown-eyed and wearing a confident smile, John Von Neumann had mastered calculus at the age of 8. As an adult, he co-invented the electronic computer and later collaborated on the Manhattan Project. But it was his work on game theory – beginning with this first lecture – that would influence the fields of economics, business, biology, and animal behavior. That lecture, on the theory of parlor games, was the precedent for modern game theory, which comprises the mathematical analysis of games. Neumann conjectured that apparently trivial decisions while playing games could be used to elucidate behaviors more widely. Von Neumann wasn't the first person to put forth the basic concepts of game theory, but like Darwin before him, he was the catalyst for a groundbreaking theory, and the person who presented the ideas in the most eloquent package.

Parlor games such as "rock, paper, scissors," provided a readily accessible vehicle for people to understand the basic concepts of game theory, which analyzes strategies during competitive situations, wherein the outcome of one player's actions is critically dependent upon the other player's actions. Rock, paper, scissors is a two-player game with very simple rules: rock beats scissors, scissors beats paper, paper beats rock. Each player puts their hands behind their back and then, on the count of three, uses their hand to make the configuration of one of the three possibilities, with a flat hand indicating paper, a fist indicating rock, and the pointer and index finger splayed indicating scissors. Players will watch their rivals closely; if they see patterns (for example, if the opponent always plays paper after rock or always plays scissors twice in a row), then they will modify their own behavior accordingly. This seemingly simple game is dependent upon the responses of the opponent; players who lose a round are more likely to change their throw during the next round than players who win. The "safest" strategy for the game, which results in a tie, is for each player to play each possibility one-third of the time, and at random, avoiding a pattern that the opponent will pick up on. Seemingly simple games like these provided the foundation for later scientists to explore game theory in even more applications.

In 1951, with far less confidence and far more eccentricity than Von Neumann, American mathematician John Forbes Nash, Jr., presented his own findings on game theory: the Nash equilibrium. Handsome, unassuming, and with side swept brown bangs, 23-year-old Nash had earned his PhD the prior year, focusing on non-cooperative games (Fig. 34). Nash, who would later be portrayed by actor Russell Crowe in the film *A Beautiful Mind*, was not yet plagued by the mental illness depicted in the film; this wouldn't arise – at least, not obviously – until some eight years later. In game theory, the Nash equilibrium is a solution to non-cooperative games with two or more players. Each player is assumed to be familiar with the other players' equilibrium strategies, and player won't experience a gain if they change their own strategy. Game theory comprises mutual outguessing, where players' strategies depend upon the strategies of their rivals. If each player has a chosen strategy and no player benefits by altering this strategy while others maintain their own, the set of strategy choices and corresponding payoffs constitutes a Nash equilibrium. For example, two dogs, Rover and Rex, would be in a Nash Equilibrium if Rover made the best choice that he could while taking into account Rex's choice, but Rex's decision was unchanged, and Rex made the best choice that he could, while taking into account Rover's choice, but Rover's decision remained unchanged. Thus, both dogs would be considering the prior actions of the other dog, and acting accordingly, but neither dog would be changing their own choices. Conditional interactions (what I do depends on what you do) are a big component of both human and canine behavior, even if these choices are occurring on an unspoken or subconscious level. So can dogs be in a Nash Equilibrium, either with other dogs or with humans, or is this more anthropomorphic than scientific? Given their history of coevolution with humans, it's possible that they could be with either dogs or with humans.

FIG. 34

John Forbes Nash. *Sketch by Emilia Galletto.*

THE PRISONER'S DILEMMA

While he was formalizing his theory of evolution, Charles Darwin realized that there was a problem with "altruistic" behavior. Given that the basis of selection occurs on the individual level, altruism, which is the act of increasing another's fitness at a cost to your own, would be counterproductive. Altruism has been observed in a wide range of species, however, leading scientists to search for explanations of this behavior, including kin selection and reciprocal altruism. Now if you're a member of social hymenoptera, an order of over 150,000 insects, including ants, bees, and wasps, (which you probably aren't, if you're reading this book), kin selection is a prudent strategy because you have a high degree of relatedness with your sisters, while other insects might not share that same degree of relatedness with their group members. Thus, kin selection makes sense for one individual to make "small sacrifices," like not bearing offspring of their own, when it means that it benefits the family unit. Consider wolves, for example: a mother and father might be the only members in the family who are reproducing, but siblings who are still in the pack and have not left yet to start their own families will aid the dependent pups with feeding, protection, company, and guidance. They're helping the family as a whole, even if they're expending their own calories or potentially delaying their own reproductive careers.

But what explains behaviors that help other individuals among non-related animals? Altruism, also known as "pro-social behavior," can be detrimental to the helper's fitness while it supports the fitness of the other individual – so given the relative "selfishness" of most individuals, why would one do this? Cases of true altruism occur with humans, but appear to be less common among nonhuman animals, with explanations ranging from mistaken genetics (the individuals were probably related) to mental shortcomings on the part of the assisting individual. When Zombi, a pregnant female chimpanzee at the Monarto Zoo in South Adelaide, Australia, adopted an orphaned chimpanzee infant named Boon, people took notice. The behavior was considered to be "unprecedented" among unrelated individuals, and the act of caring for an unrelated baby in addition to her own biological one definitely would have come at a high cost for Zombi. Scientists wondered: Would Zombi continue to care for Boon after the birth of her own offspring? Sadly, Boon passed away shortly before Zombi gave birth to her own infant, leaving this question unanswered. While rare, this altruistic behavior isn't unprecedented. A recent study found that there have been at least 18 cases of orphaned chimpanzees being adopted in the wild.[125] One of the most well-known of these instances occurred in the Ivory Coast's Taï forest, where an adult male named Freddy adopted an orphaned chimpanzee named Victor. These acts of adoption were all noteworthy because altruism has long been upheld as a uniquely human trait, much like culture and tool-making once were. But the more that we learn about nonhuman animal behavior, the more continuity of behavior we discover.

---

[125] Boesch, C., Bolé, C., Eckhardt, N., & Boesch, H. *Altruism in forest chimpanzees: The case of adoption.* http://journals.plos.org/plosone/article?id=10.1371/journal.pone.0008901.

Primates aren't alone in their altruistic behaviors, though; unsurprisingly, man's closest companion exhibits hints of altruism, as well. A recent study conducted at the Messerli Research Institute in Vienna revealed that dogs were willing to help other dogs without the expectation of reciprocation. The study used eight donor dogs and eight recipient dogs. The donors were taught to pull a rope that would then deliver a treat on a tray, and they could choose whether they would receive the treat or if they'd give it to the recipient. Rather than always choosing to feed themselves, donor dogs would often choose the empty tray and allow the recipient dog to have the treat instead. This occurred most frequently when the donor and the recipient knew each other well.[126] Other species of animals have also exhibited apparently altruistic behaviors, including vampire bats (*Desmodus rotundus*) that donate some of the blood that they harvested to those who were unsuccessful in attaining blood and vervet monkeys that give out alarm calls, even though it places themselves at risk of predation. So what would inspire behaviors such as these? While kin selection can explain the altruistic behaviors of related individuals, group selection can help explain the altruistic behaviors of non-related group members. According to group selection theory, group members are more likely to survive if they cooperate versus if they don't, so helping an unrelated group member can indirectly help oneself (and one's family, too.)

There's a popular scenario that can also help explain the phenomena of apparently altruistic behaviors. Let's say that there are two suspects in a bank robbery that are being detained at a police station. They're being questioned by the authorities in separate rooms. The authorities know that the suspects are guilty of the crime, but they don't have enough corroborating evidence to make the charges stick. Each of the suspects is being offered a plea deal – if and only if he provides the authorities with information about the other suspect. The suspects weigh their options, trying to decide whether it's "worth it" to defect and turn in their co-conspirator, or whether it's potentially too costly. But this isn't just a plot device for a popular crime program; it's also an evolutionary reality that's referred to as the "prisoner's dilemma" that demonstrates the complexities of cooperative behavior.

W.D. Hamilton and social scientist Robert Axelrod studied the evolution of cooperation, which is illustrated by game theory and by a famous game called the Prisoner's Dilemma, or PD. In the PD, the authorities have just enough evidence against each of two suspects to convict them for a number of years (referred to as "R" years). If both of the suspects choose to provide evidence against one another, then they will both be convicted and sent to jail for P years, with $P > R$. Given this formula, neither suspect would ever feel compelled to provide evidence against the other, but in the PD, much like popular crime shows, the suspects are interviewed separately and without the knowledge of the other suspect to have the opportunity of a comparatively protracted sentence (referred to as T, with $T < R$), if he will provide the authorities with evidence against the other criminal. The other criminal, in turn, would then receive a much harsher sentence, referred to as S, with $S > P$. The PD can

---

[126] Quervel-Chaumette, M., Dale, R., Marshall-Pescini, S., & Range, F. (2005). Familiarity affects other-regarding preferences in pet dogs. *Scientific Reports*, 5(18102).

also be applied to animal cooperation if these payoffs were denoted as "positive fitness increments" rather than jail sentences. And if animals played the PD multiple times, the benefits could outweigh the costs if they cooperated with one another.

Axelrod's 1984 treatise on cooperative behavior, *The Evolution of Cooperation*, asks why and how humans choose to work with one another for mutual benefit – and it was the catalyst for a new industry of research. Axelrod begins with the question that has been asked for centuries: "Under what conditions will cooperation emerge in a world of egoists without central authority?" In a two-player, non-zero-sum game, "row" players are pitted against "column" players. He proffers the Prisoner's Dilemma as the basis for his entire book. Axelrod writes:

> *"One player chooses a row, either cooperating or defecting. The other player simultaneously chooses a column, either cooperating or defecting. Together, these choices result in one of the four possible outcomes shown in that matrix. If both players cooperate, both do fairly well. Both get R, the reward for mutual cooperation, In the concrete illustration of figure 1 the reward is 3 points. This number might, for example, be a payoff in dollars that each player gets for that outcome. If one player cooperates but the other defects, the defecting player gets the temptation to defect, while the cooperating player gets the sucker's payoff. In the example, these are 5 points and 0 points respectively. If both defect, both get 1 point, the punishment for mutual defection."*

Players can cooperate, receiving 3 points of the reward for their reciprocal support, or they can defect and potentially get 5 points – the "temptation payoff." But if you think that the other player will also defect, you're faced with the chance that you'll receive 1 point if you also defect or 0 points if you choose to cooperate. Thus, defection is the better strategy if you believe that the other player will cooperate, and it's better to defect if you think that the other player will defect, as well.

"But," Axelrod warns, "the same logic holds true for the other player, as well." So what's a wise player to do? Well, other players would defect, regardless of what they expect you to do, so then both players should defect. But this leads to a worse outcome for both players: each receives only one point instead of the three points each would receive for mutual cooperation. That's where the dilemma comes in. If it's a noniterated (one-time-only) game, then it's a relatively small loss, but over multiple iterations, the difference between 1 and 3 begins to add up.

According to Axelrod, the Prisoner's Dilemma has a second part: the players can't alternate between exploiting one another to get out of the dilemma. This changes everything, because with an indefinite number of interactions, that's where cooperation can emerge. That difference between 1 and 3 becomes significant when there's a long-term relationship in play; rewards for mutually cooperating will outweigh the temptation to defect. Iterated interactions provide the necessary and sufficient conditions for cooperative behavior to emerge. A zero-sum game is the belief that a win for one side is a loss for the other, and vice versa, while a non zero-sum game involves the cooperation of both sides for a win-win situation. It is with non zero-sum games where trust can evolve.

What's the difference between iterated games and non-iterated ones – and why is this difference so important? Iterated games occur repeatedly, while non-iterated games – one-offs – are just that; one-time-only occurrences wherein there's little risk to either party if they're greedy and "defect." Over the course of an iterated game, players can develop strategies contingent upon the other player's past moves, either being generous or retributive.

So how is this represented in possible "real world" applications? The game has to be two-way interactions between specific individuals. Models of this game have created different "character" strategists for the Prisoner's Dilemma, including "copycat," "grudger," "always cheat," "always cooperate," and "cooperate." The copycat, also referred to as the Golden Rule, Tit for Tat, and reciprocal altruism, always starts with "cooperate" for their initial move and then just copies what their partner does. The grudger starts with cooperation, but if the other player cheats, the grudger will cheat indefinitely. The detective alternates between cooperating and cheating, acting like the copycat if the other player cheats in return and playing like the "always cheat" if the other player never cheats. So how do you determine the "best" strategy? In the 1970s, Axelrod held a tournament pitting the various strategies against one another. Under multiple iterations of this game (more than five, for example), the Copycat/Tit For Tat strategy was the most successful – and many nonhuman animals appear to think so, too.

The stickleback fish (*Gasterosteus aculeatus*) inhabits inland coastal waters and uses a copycat strategy when approaching predators. This species usually pairs off with another fish, and then the two partners swim side by side (Fig. 35). When presented with a predator, stickleback fish in the wild would use a Tit for Tat strategy to inspect the potential threat. The first fish would surge ahead by a length or two to inspect the predator, and the second would cooperate by following closely afterward

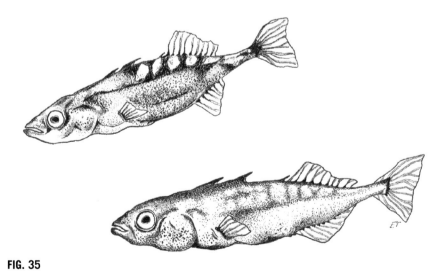

FIG. 35

Stickleback fish. *Sketch by Emilia Galletto.*

and also inspecting the predator. The possible cost of defection was high: the predator could eat one or both of the fish. In a controlled experiment, the first fish would stay back if a simulated conspecific also held back, and if the simulated conspecific went forward, the fish would reciprocate. In areas that had higher predation risk, the fish tended to be more cooperative, as well.

TO TWEET OR NOT TO TWEET?

For some birds, that is the question. In the Eastern and Midwestern United States, an 11-gram songbird with a striking black hood, a yellow face, and a yellow underbelly sings out to an unfamiliar neighboring bird, who returns his call. The songs of the two birds fill the air as they call back and forth to one another, their chests puffed out, their wings half-raised, and their heads craning toward the sound of the unfamiliar avian. When a third bird, a neighbor who is familiar to the first one, also sings out, the first bird does not respond. And Tit for Tat explains this. Hooded warblers (*Wilsonia citrina*) are a territorial songbird species that uses a "dear enemy" relationship with neighboring birds (Fig. 36). Males of this species recognize the calls of neighbors

FIG. 36

The hooded warbler. *Sketch by Emilia Galletto.*

and reduce their responses to them, as a non-response requires less energy, and this can be mutually beneficial for both neighbors. The dear enemy effect will occur when two neighboring territorial animals exhibit decreasing levels of aggression after their territorial demarcations are made. The familiar neighboring birds no longer need to call out to tell one another where the "property lines" are, while the unfamiliar bird will need to be informed that he's in someone else's territory. In effect, good fences (or good songs) make for (better) neighbors. As a result of this familiarity, the males no longer need to defend against them; these known neighbors aren't allies, but they aren't strangers, either. In the presence of an unfamiliar or new neighbor, however, defense of territory behaviors ramp up once again. The dear enemy effect is an evolutionarily stable strategy when the response reduction is reciprocal and is the converse of the nasty neighbour effect, where species exhibit more aggression toward familiar neighbors than to unfamiliar ones. Similarly, when a dog is patrolling their fence line, they might exhibit fewer defense of territory behaviors (barking, growling, charging at the fence) with a familiar (but not necessarily "friendly") neighbour dog than with an unfamiliar dog who is walking past their property. This is, of course, dependent upon breed and individual personality; dogs who have been bred for generations to be guard animals will likely have far different reactions than, say, a teacup poodle.

So how are these animals making these choices, based upon the behaviors of another? Tit for Tat has three underlying ideas to its strategy. The initial move with this strategy is to cooperate. The first idea is that if the partner defects after the first player first cooperates, they will reciprocate with retaliatory defection. The second idea is that players need to demonstrate forgiveness, wherein a player forgives a player that showed a prior defection if they have subsequently cooperated. The third idea behind the Tit for Tat component is being nice, wherein both players' opening moves are cooperation. The most successful strategy for Tit for Tat is using the "nice" component, demonstrating that cooperation can be mutually lucrative.

When there's a lower win-win reward, "always cheat" tends to win, because there isn't as much of an incentive to cooperate. Tit For Tat is a safe, conservative strategy, but if two copycats are playing with one another, miscommunication or mistakes can lead to retaliation. That miscommunication can then lead to the failure of both players.

Additional "character" strategists for the Prisoner's Dilemma include the Copykitten, who cheats back only after the other player has cheated twice in a row, the Simpleton, who starts with cooperation, and then if the other player cooperates in turn, will do the same thing, even if it was a mistake; if the other player cheats back, the Simpleton does the opposite thing as the last move. And finally, there's the Random player, which cheats or cooperates randomly, irrespective of the other player's prior move.

Game theory shows us that iterated games, with the knowledge of the other player's possible future plays, are required for trust to evolve. For a win-win to occur, the players must be playing a non-zero-sum game. And if there's a high rate of miscommunication, there's also a rapid degradation of trust; miscommunication rates of 50% or higher result in a loss for all players.

EQUAL PAY FOR EQUAL WORK

Domestic dogs provide a unique way to explore questions of inequity. Are our dogs aware of what might be "fair?" They're highly social beings and appear to abide by certain "rules" when it comes to play, but can they also detect when there's an inequity between two individuals? Some studies appear to indicate that dogs, like people, can tell when they're getting short-changed.

In an alternative scenario called an "inequitable payoff," two people are given the opportunity to share some found money – let's say one dollar. The first player (the proposer) chooses how the money will be divvied up, while the second player (the responder) either accepts or rejects this division. If the responder rejects the proposer's offer, then neither person receives any money, but if the responder agrees, the money is divvied up as the first player has determined. From a purely game theory point of view, the "best" strategy for the proposer would be to offer the minimal amount that the other person would accept, which would be one penny, because one penny is the smallest amount that's larger than nothing. But when the proposer offers the responder this amount of money, rather than accepting it so that both of them will receive money, the responder declines. When the responder views the division as inequitable, they will decline, even if that means that they both receive nothing. There are actually two kinds of payoff at play here: the financial portion of it and the fairness portion. When the division isn't equal (or close to equal), the responder would rather deprive the proposer of the money than receive a small benefit of their own. Playing the game many times in sequence, players can adopt "punishment" strategies. Both players can learn "don't do that," however, and learn how they can both benefit.

Humans aren't the only species that takes umbrage to inequitable payoffs. Experiments with multiple non-human primate species, including the brown capuchin monkey (*Cebus apella*) revealed that monkeys refused to participate in a social experiment if they viewed a peer receiving a more valuable reward (in this instance, grapes instead of cucumbers) for performing the same task that they did. The study suggested that inequity aversion likely had a very early evolutionary origin.[127] Footage of experiments such as these depicts the monkey who is receiving the inferior reward slapping the ground with their hands, hitting the caging, and showing their teeth. You don't need to be a primatologist to get the gist that this isn't a happy monkey, and it might even remind you of a time when you discovered that a peer was earning more money than you were for performing the same work (or less). The "cheated" monkeys' reactions were amusing because human observers could see so much of themselves in them. Further studies revealed that monkeys who saw their peers receive a more valuable reward than themselves also put forth the least amount of effort in their own tasks.[128] There were similar responses to inequitable payoff studies with chimpanzees, tamarins, wolves, and dogs.

---

[127] Brosnan, S. F., & de Waal, F. B. M. (2003). Monkeys reject unequal pay. *Nature*, *425*, 18.
[128] van Wolkenten, M., Brosnan, S. F., & de Waal, F. B. M. (2007). Inequity responses of monkeys modified by effort. *PNAS*, *104*, 47.

Studies with dogs have provided support for inequity aversion in this species. One study found that the absence of a reward reduces inequity aversion in dogs,[129] with dogs changing their behavior when their partner received a food reward. Unlike the results of primate studies, however, the dogs didn't respond to qualitative differences with the food or the effort. A second study revealed that dogs would preferentially work with trainers who over-rewarded over those who would provide "fair" recompense, indicating that dogs have at least a precursory awareness to inequity.[130]

Dogs' ability to detect inequitable outcomes had been attributed to domestication, but in a recent study, wolves demonstrated similar abilities. Dogs and wolves were tested in adjacent enclosures where they had to press a button with their paws to receive a reward. In the no-reward condition, the first canine received an award, while the second did not. In the quality condition, both received an award, but the higher valued reward was again given to the first canine. The second canine, much like the capuchin monkeys, began to refuse to participate. Interestingly, wolves appeared to hold a grudge against the experimenters when they were the ones who did not receive an award: they acted more aloof from the humans than before the experiment, while the domesticated dogs did not.[131]

HOW I BEHAVE DEPENDS UPON HOW YOU'RE BEHAVING... MAYBE!

Rock, paper, scissors, the prisoner's dilemma, and inequity aversion all use the adage, "How I behave depends upon how you are behaving... (maybe!)" Game theory, which is the study of mathematical models of conflict and cooperation, has been applied to political science, war, logic, computer science, economics, business, psychology, and most recently, to biology. While game theory originally addressed zero-sum games wherein the loss of one resulted in the gain of another, it now has wide-ranging applications for animal behavior.

The concepts of game theory that are so fundamental to modern animal behavior science include conditional costs and benefits (my costs and benefits of being social are relative to your costs and benefits of being social), rules of thumb, and iterated versus non-iterated interactions, which are directly applicable to practical dog behavior, especially in cases of social behavior issues. They also make perfect sense to owners: We all experience these factors in our everyday interactions, and when explained to owners from the perspective of their dogs, it makes sense.

---

[129] Range, F., Horn, L., Viranyi, Z., & Huber, L. (2009). The absence of reward induces inequity aversion in dogs. *PNAS, 106*(1), 340–345. https://doi.org/10.1073/pnas.0810957105.

[130] Horowitz, S. (2012). Fair is fine, but more is better: Limits to inequity aversion in the domestic dog. *Social Justice Research, 25*(2), 195–212.

[131] Essler, J. L., Marshall-Pescini, S., & Range, F. (2017). Domestication does not explain the presence of inequity aversion in dogs. *Current Biology*.

Evolutionary game theory (EGT) applies game theory to evolving populations; it's the perfect marriage of economics and ecology. EGT was pioneered in 1973 when theoretical mathematician and evolutionary biologist John Maynard Smith realized that players didn't necessarily have to behave rationally; they only needed to have a strategy. Maynard, along with population geneticist George R. Price, explored this by formalizing contests, analyzed as strategies, and the mathematical criteria that could be used to predict the results of competing strategies. EGT can be used to explain a wide range of behaviors, including apparently altruistic ones. Routes to apparent "altruism" include kin selection, indirect reciprocity, direct reciprocity, and reciprocal altruism – and all of these strategies can be explained mathematically with EGT. In EGT, strategies aren't the deliberate actions of two or more competitors, but inheritable traits; and the payoffs are Darwinian fitness, measured by average reproductive success. The "players" in EGT are members of a population who are all competing to pass their genes on and have a larger share of descendants than their rivals. Unlike classical game theory, players in EGT are born with a strategy; their offspring later follow this same strategy, as well.

The side-botched lizard, *Uta stansburiana*, provides a biological example of the strategies exhibited in rock, paper, scissors (Fig. 37). This lizard is a polymorph with three subtypes, each of which follows a different mating strategy. The first morph is an orange-throated male that exhibits high levels of aggression and oversees a large territory, within which he attempts to mate with multiple females. The second morph is a yellow-throated male that exhibits low levels of aggression, but mimics the appearance and behavior of female lizards. This morph tries to sneak into the territory of the orange-throated male to mate with the females, thus taking over the population. The third morph is a blue-throated male that mates with and carefully guards

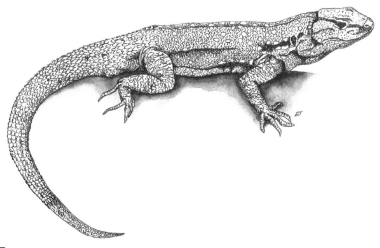

FIG. 37

The side-botched lizard. *Sketch by Emilia Galletto.*

a lone female; because of his vigilance, sneaky, yellow-throated males cannot mate with this female. Blue throated lizards also can't compensate for the aggression of orange throated ones. Thus, this species has a six-year population cycle that corresponds closely to rock, paper, scissors.

Food, territory, and mates are all potentially shareable resources. Ethologists Konrad Lorenz and Niko Tinbergen were searching for a way to resolve contests over shareable resources. In response, John Maynard Smith analyzed a scenario called the Hawk-Dove game. This game involves a polymorph species with two subtypes, each with their own strategies. The hawk's strategy is to first display aggression, escalating into a fight until it wins or is injured. The dove's strategy is to first display aggression, but if he is faced with an escalating situation, he flees rather than fights. But if there isn't escalation, the dove will attempt to share the resource. If most of the population consists of doves, the hawk subtype will proliferate, but if most of the population are hawks, the escalating conflict will lead with a probability of 1/2 to injury.

Evolutionarily stable strategies (ESS) such as the hawk-dove game aren't optimal, nor are they singular solutions. An ESS isn't always present, either; for example, the side-botched lizard scenario is a rock, paper, scissors evolutionary game with no ESS. ESS is also not an unbeatable strategy; it's only uninvadable. For example, if a population adopts a certain ESS, that environment can't be invaded by alternative strategies that are initially rare.

Game theory can be used to explain many of the "unexplained" or "unexpected" behaviors that we witness with our dogs. The signs to watch for when diagnosing social issues in dogs, whether dog-human or dog-dog, are whether the behaviors change when in proximity to other animals, and even more specifically, whether they change differently when in proximity to different animals. That is, dogs are constantly evaluating not only the costs and benefits of their decisions, but the *environment* in which those costs and benefits are occurring. The best decision for me may, and often does, depend on the decision that you make. It can all become very complex, and especially in intense, fast-moving situations or situations in which information may not be communicated clearly, mistakes can be made. And mistakes may have negative results, which result in a learning event, leading to changes in future evaluations of costs and benefits ("I remember what happened last time…"), and so on and so on, down a rabbit hole of deteriorating behavior.

There is a second principle inherent to conditional costs and benefits, and it's well illustrated by game-theory approaches: iterated versus noniterated interactions. Decades of game-theoretical research has shown us that optimal decisions may frequently be influenced by whether you ever expect to see those social interactors again. Think about it: if you expect to interact with someone again in the near future, you frequently behave differently towards them. This is frequently seen in wait-staff tips: if you'll return to the restaurant, you tip better. Regulars at the local bar tip better, and regular customers at the grocery checkout are friendlier to the staff than the out-of-towners passing through.

This is because of the phenomenon of reciprocal altruism: "If you scratch my back now, I'll scratch your back later." But this is all based upon the fact that I will be *around* later to pay back the favor. This concept drives much of our social behavior: It is the underpinnings of cooperation. And it drives our dogs' behavior, too: So many issues in dog behavior can be attributed, at least in part, to stability of the social environment. And game theory, differences between iterated and non-iterated situations, and reciprocal altruism all help explain these effects.

Game theory does have specific applications for the dog world, particularly with conditional interactions and with the importance of iterated versus non-iterated games. Dogs housed within animal shelters and dogs whose owners foster a lot of other dogs will often exhibit aggressive and stress behaviors. This is a clear non-iterated situation, as social interactors are constantly changing. Dogs, like humans, will think, "If I am going to interact with you again and again, I will behave in one way, but if I will never see you again – probably – I will behave differently."

The dogs can't count on reciprocation, so they decide to "take what they can get when they can get it." Clients often ask JCH, "Why does my dog behave differently at the off-leash dog park than at dog daycare?" Well, it depends on how socially stable the dog park is. A small-town, same time of day every time scenario would have the same dogs present during most social interactions. This is an iterated "game," and it's a great situation. In a big city, dog park, where the dog is visiting at different times and days, that's always a different social environment. This is a non-iterated social environment, and often causes stress and aggression among dogs, as it's a very different evaluation of the costs and benefits to interactions.

In practical applications, this is important in cases where there are a lot of transient dogs: a classic case is stress-related behaviors in a resident dog in a home which does a lot of fostering. Fostering entails having a lot of transient, passing through dogs and at times, can be VERY difficult for a resident dog. JCH has dealt with several cases where the owner had to stop fostering or at least provide the resident dog with his own space. SOME dogs don't care either way, some dogs love having a variety of dogs stay in their homes with them, and some dogs don't like it at all: and that's personality!

When TLC found a dog in her neighborhood who had been missing for more than a week, she brought her home. The owner of the dog was out of state, so the dog, Nutmeg, had to stay at TLC's house overnight. TLC stayed in the spare room with the new dog, who was exhibiting signs of anxiety (shaking, breathing heavily, yawning frequently, whining). Jack became anxious at the presence of the new dog and began to growl outside of the spare room door, a behavior that was very unusual for him. Jack continued to act strangely for several days after Nutmeg left, indicating that he wouldn't like to have unfamiliar dogs in his home.

The conditional game and the iterated game are both very important in shelters. In the dog's mind, it might be something like: there's so much transience that I can't learn how you are going to behave, so I don't know how to best play my game (all unconscious of course), and since I quickly learn that you guys come and go, I do best by defecting: not cooperating or playing nice...an inherent, fundamental issue in shelters! But of course, not literally...evolution has shaped this into: unless I see you again and again, then I defect...

These situations bring together so many of the principles we have discussed: communication, and lack of same among some breeds; costs and benefits, as perceived in the dog's world and influenced by her personality and learned responses; social behavior, which is so important to a cooperative species like dogs and conditional upon the decisions of others; and opportunities for iterated reciprocity, or not in the large, ever-changing city populations in which we have recently placed many of our pets.

As JCH diagnoses behavior issues, these are the factors he needs to know: personality, social environment, learning history (like trauma or shelter experience), and how these factors have come together in the dog's mind. JCH was working with a dog named Carter, a beach dog from Costa Rica who had been "adopted" and brought home by newlyweds on their honeymoon. Prior to this adoption, Carter had learned to survive in ever-changing, unstable social environments on the streets and beaches of a small town in Costa Rica. Any attempt at cooperating or sharing was quickly disabused in his early life. This had become locked in to his behavior because of the early exposure to such an environment. In addition, he had health issues. When he was brought to the United States, he was kenneled for weeks for quarantine, and then longer for health treatment. When he was finally at home, he was distant and antisocial with humans unless they had food in their hand, and he was very competitive with other dogs.

While displeasing to his owners, Carter's behaviors weren't unusual, given his prior social experiences. There were so many issues, and Carter's adopters were passionate about helping him, and yet, JCH's very first advice was, "Do nothing. Leave him alone...feed him, keep him warm, give him a protected private space, and just let him chill...for perhaps weeks. Don't try to interact, but interact if he seeks it, quietly and briefly. Let him have some control of his life back, and most importantly, let him learn that he was now in an iterated, stable situation. Let him learn that he has predictable food, predictable social interactions, and predictable, stable life."

It took six weeks of nothing more than stability, for (perhaps the real) Carter to emerge: a very social, playful pup who required some obedience training and some polite manners training, but whose aggression and competitiveness disappeared entirely. JCH finds that this same phenomenon occurs frequently in dogs adopted from shelters, albeit in usually a less dramatic form.

Game theory can not only help us understand why our dogs make the decisions that they do; it can help us predict future behaviors, as well. This was demonstrated when novel behaviors were observed with ravens. Ravens typically search for food individually, with younger birds often "recruiting" the help or protection of older group members. In 2009, though, juvenile ravens in North Wales were observed foraging for food in "gangs," where they used these behaviors to both acquire food and to attain higher social status.[132] The ravens were exhibiting a tradeoff between foraging efficiency and opportunities for social climbing – a behavior that had been

---

[132] Dall, S. R. X., & Wright, J. (2009). Rich pickings near large communal roosts favour 'gang' foraging by juvenile common ravens, *Corvus corax*. *PLoS ONE*.

predicted by game theory, but until that point, had been unseen. Understanding how dogs view inequity, iterative versus non-iterative games, and conditional costs and benefits can help us improve their quality of life and our social relationships with them.

CANINE BRAINPOWER

Prior to the groundbreaking research of the mid-twentieth century ethologists, many behaviors were considered to be solely within the domain of humans. Those behaviors included components of "Theory of Mind (ToM)," including perspective taking and deception. In recent years, however, a wide range of animals have been shown to exhibit these capabilities. Deception in particular has been upheld as an important aspect of theory of mind because it requires an individual to understand that there is a perspective other than their own and that this perspective can be manipulated to their advantage. Among stable signaling systems, deception is relatively uncommon, but it can provide a selective advantage during instances of dominant/subordinate competition and when an individual is faced with a partner that doesn't cooperate with them. As "honest" as we'd like to think our dogs are, recent research has shown that they're pretty good at being deceptive, too. A study published in *Animal Cognition*[133] provided evidence that dogs can act deceptively when they are faced with a non-cooperative partner. During a training routine, dogs were faced with one of three circumstances to acquire food: an always-cooperative owner, a cooperative stranger, and a competitive, uncooperating stranger. The dog could lead his partner to one of three locations: one with no food, one with a "boring" food option (dry food pellets), and the last with a highly favored food (sausages). Dogs led their cooperative partners to the sausage location far more often than the dry pellet and non-food options, while they led their competitive partners to the location with no food more often than the two locations with food. Thus, while the dogs knew where the desirable food was, they demonstrated tactical deception when faced with a partner that wouldn't reciprocate.

This demonstrates the complexity of canine cognition and behavior. The ramification is now that we know that our dogs are capable of such behavior, how is it being used in our interactions? Dog behavior is far more complicated than most people think!

Dogs won't only be deceptive depending upon whether their partner will cooperate or compete, though; they'll also react differently depending upon what they know another individual can sense. Recall Jack's food-stealing choice. He was taking a risk that he believed to be worth it, given the strong chance that he would be reprimanded and the slight chance that he might get sick from the food. But he wasn't just taking a risk; he was also doing this with extreme stealth. And Jack isn't the only dog who has behaved in a way that demonstrates an understanding of what others can see or

---

[133] Heberlein, M. T. E., Manser, M. B., & Turner, D. C. (2017). Deceptive-like behaviour in dogs (*Canis familiaris*). *Animal Cognition*.

hear. Studies have shown that a wide range of primate species will take into account what another individual can hear during a competitive situation. Dogs will also react differently in a situation depending upon whether they think someone can hear what they're doing. In a study published in *Applied Animal Behaviour*,[134] dogs were commanded to not take a treat out of one of two containers. A human "gatekeeper" was either oriented toward them or oriented away. The first container was silent when the food was taken in or out of it, thus allowing for stealth, while the second container would make a noise. The study found that "…dogs would preferentially attempt to retrieve their food silently, but only when silence was germane to obtaining food unobserved by a human 'gatekeeper'."

While this is a very preliminary study, it provides further evidence that dogs use Theory of Mind (ToM) with humans to choose which site from which to remove food, based upon auditory ToM. If the human was watching, it didn't matter which, but if the human was oriented away, the dogs used the silent option. This study was also species-specific; most studies are oriented toward the visual modality, as primates (including humans) are highly visual beings. Dogs, however, rely more upon their sense of olfaction and hearing than they do on their sense of sight, so it's important to not only ask the right questions (such as can our dogs tell what we're able to perceive?) but also to ask them in the right way, using modalities that are most salient to them.

With their history of human co-evolution, it shouldn't be surprising that dogs actually perform better than humans' ape cousins with tasks such as referential pointing, indicating that dogs know to look in the direction that the human is pointing. It isn't just nonhuman apes who are outperformed by dogs with referential pointing; other domesticated species, such as horses and cats, don't fare as well in this task, either. Somewhere during their co-evolution, referential pointing was particularly salient for dogs – it was a silent mode of interspecies communication. Referential pointing is typically tested using an object-choice task, where dogs would follow the direction that a person pointed. This experimental paradigm is typically used to find a desired food item. A recent meta-analysis found that, across all dog breeds, there was a shared ability to understand referential pointing. This indicates that understanding why a human was pointing was a fairly early adaptation for domesticated dogs.[135]

---

[134] Kundeya, S. M. A., De Los Reyes, A., Taglanga, C., Allena, R., Molina, S., Royer, E., et al. (2010). Domesticated dogs (*Canis familiaris*) react to what others can and cannot hear. *Applied Animal Behaviour Science, 126*, 45–50.

[135] Dorey, N. R., Udell, M. A. R., & Wynne, C. D. L. (2009). Breed differences in dogs sensitivity to human points: A meta-analysis. *Behavioural Processes, 81*, 409–415.

CHAPTER 9

Debunking dominance: Canine social structure and behavioral ecology

THE SOCIOECOLOGY OF SOCIAL BEHAVIOR

The social structure of a species, and hence their social behavior, is based upon resource distribution. When resources are restricted, a nonviolent method for distribution is needed, and different species have chosen different strategies for this. While bonobos and chimpanzees only differ in their DNA by .4%,[136] the same genetic difference that demarcates domesticated dogs share with wolves, they exhibit vast differences in their social structure and in their behavior. Bonobos are female-dominant; females form alliances against the males in the group through same-sex socio-sexual activity that is hypothesized to diminish stress levels. Bonobos allow conspecifics from neighboring groups to overlap into their territories and have been seen mating with outside individuals. Chimpanzees, conversely, are male-dominant, exhibiting aggression (often lethal aggression) to outside chimpanzees who enter into their territory. Chimpanzees will aggressively patrol the boundaries of their territory, which is specific to each social group. Bonobos have not been witnessed hunting or exhibiting aggression in free-living situations, while chimpanzees frequently have coordinated hunting parties wherein members have specific "roles" while they hunt and consume monkeys (and sometimes, other chimpanzees, too.) Bonobos often use sexual behavior for social bonding, tension reduction, greeting, conflict resolution, and to elicit food or social benefits from the recipient. Among chimpanzees, the highest-ranking males monopolize the females who are in estrus, increasing the likelihood that they will father their offspring. Two species, just as closely related as wolves are to domesticated dogs, and with just as many behavioral differences (Fig. 38).

One hypothesis for the significant differences between the two species is their historically variant ecological factors, including the availability of food resources such as terrestrial herbs, the distribution of fruit, including fruit patch size, and the proximity between distributed resources. Chimpanzee societies exhibit a fission-fusion system, wherein members of a family group will break up during the day and then rejoin in the evening. Females and their dependent offspring will often break off in separate groups, as will males, who often join hunting parties. During

---

[136] www.sciencemag.org/news/2012/06/bonobos-join-chimps-closest-human-relatives.

FIG. 38

Female chimpanzees and their dependent offspring, like this infant, break up from the rest of their family group during the day and rejoin them in the evening. *Sketch by Arianne Taylor.*

the Pleistocene, chimpanzee populations experienced dry periods that restricted food resources and there were likely selective pressures for this social system. Bonobos, however, benefited from a refugia in their territory where resources were not limited; food was more readily available and more abundant. With this even resource distribution, all of the group members, including the females, did not have to incur the costs of scramble and contest food competition.[137]

Scramble competition occurs when a resource is not monopolizable by one individual or one group, but is instead accessible to all who are seeking it. The resource is typically finite, resulting in the first individuals to arrive reaping most of the resource. Contest competition occurs when two or more individuals compete for a resource that is monopolizable, such as food or territory. When resources are limited, hierarchies are one way to resolve this issue. Species with clear dominance hierarchies include baboons, macaques, chimpanzees, humans, lions, wolves, and dogs. Attaining the highest rank can be based upon a multitude of factors, including coercive behavior, physical altercations, and superior social skills. When resources aren't limited, though, as in the case of the bonobos, there's no need for a strict dominance hierarchy.

The divergent socioecology of wolves and dogs has dramatically impacted their social behavior, as well. While domestication fostered a more human-friendly temperament in dogs, they also became more dependent upon their human family members. With a clearly established familial dominance hierarchy, intra-group

---

[137] Furuichi, T. (2009). Factors underlying party size differences between chimpanzees and bonobos: A review and hypotheses for future study. *Primates, 50,* 197–209.

aggression among wolves is relatively infrequent. Wolves, who have a high degree of intra-group interdependence, exhibit conflict management, consolation, and reconciliation.[138] Studies examining cooperative behavior among dogs and wolves have revealed that wolves have superior cooperative behaviors with one another than dogs do with other dogs[139]; instead of relying upon the assistance of same-species dyads or triads, dogs defer to their humans for assistance. In "unsolvable task" problem-solving tests, wolves perseverate: they are far more persistent than domesticated or free-ranging feral dogs. Domesticated dogs, however, "look back" at their humans, which is considered to be a request for assistance.[140]

Our dogs ask us for assistance outside of test scenarios, as well. Jack has three legs and one eye. When he received a food puzzle that a "typical" dog would be able to solve, he struggled. This was due largely to the design of the apparatus: it was a plastic bottle with a rope to pull through the opening. Once the rope was pulled out, the treats within were accessible. This task, unfortunately, required the coordinated use of two front paws (Jack only has one) while using the mouth to pull back on the rope. After several failed attempts with this, Jack solicited the help of TLC by grabbing her hand with his mouth and placing it on the other side of the jar. Jack then held onto one side while TLC held the other; he pulled the rope, treats came out, and he was rewarded for solving the task, albeit not in the way that the manufacturer had intended.

AGGRESSION OR DOMINANCE?

It's important to denote the differences between "aggression" and "dominance" before exploring either topic, because they aren't the same thing. People toss around all kinds of terms and types of aggression, but we want to be sure to define aggression and talk about the ethology-based types of aggression, as established in the scientific literature. Aggression doesn't have an intrinsic metric, but there are observational and biochemical ways to measure aggressive behaviors. Ethology-based types of aggression include: fear or anxiety-based aggression (which is by far the most common), pain-related (which is also surprisingly common), redirected (a major issue for confrontation training methods, including dominance-based or poorly-executed aversive-based training), resource-guarding or protective (broadly defined: object, food, space [for example, territoriality]), dominance (which is rare, only exhibited in some breeds, sometimes human-directed, and usually only dog-directed), predatory

---

[138] Baana, C., Bergmüllera, R., Smith, D. W., & Molnara, B. (2014). Conflict management in free-ranging wolves, *Canis lupus. Animal Behaviour, 90*, 327–334.

[139] Marshall-Pescinia, S., Schwarza, J. F. L., Kostelnika, I., Virányia, Z., & Range, F. (2017). Importance of a species' socioecology: Wolves outperform dogs in a conspecific cooperation task. *Proceedings of the National Academy of Sciences of the United States of America*. www.pnas.org/cgi/doi/10.1073/pnas.1709027114.

[140] Marshall-Pescini, S., Rao, A., Viranyi, Z., & Range, F. (2017). The role of domestication and experience in 'looking back' towards humans in an unsolvable task. *Scientific Reports, 7*, 46636.

drive (usually with large dog/small dog problems and at off-leash parks), and male-male (exhibited by un-neutered males). Other, extremely rare types of aggression include female-female and maternal aggression. A final category of aggression would be "learned," as in trained work dogs such as K9s, who are under complete conscious control and command (Fig. 39).

While some people might say that a dog's aggressive actions were "unprovoked" or without warning, aggression, including bites, are almost always preceded by tell-tale signs. Our dogs are continuously communicating with us, via body language, vocalizations, and facial expressions. While there is individual variation and there are breed- and type-specific behavioral differences (for example, high-seek dogs that were bred for herding), dog language is universal.

Consider the physiology of anger and aggression. How do you feel when you have recently been aggressive or when you are faced with another individual who has been aggressive? Your blood pressure raises, as does your heart rate, and you might get a sudden headache or dizziness as a result. Did you have a hard time gathering your thoughts? Did your face feel hot? Did you feel shaky or panicky? Now imagine how your dog must be feeling.

The vast majority of aggressive behavior cases are due to the first two causes, and Fear-Anxiety and Redirected aggression are largely due to improper use of confrontational training methods. It is suspected that increasing levels of inbreeding, at least in the United States, is leading to increasing levels of genetically-based anxiety behavior, as well. Aggression cases require a behavior specialist to properly

FIG. 39

Darwin's hostile dog.

diagnose and treat: a Certified Applied Animal Behaviorist (CAAB) or a board-certified veterinarian (these are PhD, Master's degree, or DVM vets with extensive certified behavior training. So when do these types of aggression occur? Fear or anxiety-based aggression can happen when an individual is startled, fearful, or uncertain about a situation that they are in. Redirected aggression can occur when an individual has had an unpleasant experience and then dumps this negativity on a nearby unsuspecting victim. Resource-guarding, or protective aggression, can occur when an individual is protecting something that he or she is possessive of. Do any of these sound familiar? Humans demonstrate these forms of aggression, as well. Consider the last time that you or someone you were interacting with was aggressive – was it due to one of these reasons? If you were the passenger in a car that began to slide uncontrollably on the ice and then the driver began to yell at you, this would be fear-based aggression. If someone was reprimanded by their superior and then aggressively criticized a peer, this would be redirected aggression. And if your neighbor placed a new fence dividing your two properties, but that fence was actually over the line of demarcation, and you yelled at them, that would be resource-guarding aggression.

So, what's the difference between "aggression" and "dominance?" Aggression can be exhibited in one of the aforementioned contexts, and has both temporal and situational components to it, while dominance is a rarer occurrence and relates to an individual's impression of their social standing. Aggression is when an individual acts out in a hostile way. For dogs, aggression manifests in a variety of ways, including, but not limited to, lunging, growling, snarling, lip curling, and biting. Dominance is the rank that one has in a social hierarchy; the dominant one is at the top of the hierarchy, while the submissive one is at the bottom. Across multiple species, dominance enables group members to determine who gets access to which resources – and when – and it helps alleviate stress levels, unless you're the lowest ranking or second ranking member in the group.

Among chimpanzees, the alpha male, or most dominant member of the group, is often chosen not because he's the most aggressive individual, but because he's the most intelligent or politically savvy – he knows how to manipulate the situation. Dominance is when one individual has power or influence over others. A dominant individual *can* be aggressive from time to time, but aggression and dominance are not the same thing.

So was the new dog that Ziva met acting aggressive or dominant? Actually, the answer is neither one – the new dog was actually demonstrating behaviors indicative of insecurity. He was cowering, his tail was low, and he was exposing his teeth. Taking one of these behaviors alone might not paint a complete picture – particularly showing his teeth – but taken together, they're a clear representation of an animal who's fearful and insecure. When an animal shows their teeth, the usual implied message is, "See these weapons? I'm prepared to use them on you." But that's not always the case. Imagine putting your hand on a hot burner or seeing a car accident. You're probably going to automatically make what's called a "fear grimace": your lips are going to pull back and you're going to make a "smile" between tightly clenched teeth. But this isn't a smile or a sign of aggression; it's a fearful response, and it's exhibited across many animal species. When you see a greeting card with a "smiling" chimpanzee or orangutan on it, what you're actually seeing is a fear grimace. And, just like with humans, that fear grimace is an unconscious response in anticipation of something unpleasant. That ape was likely punished by a trainer to make that expression. A common training "trick" is to put a metal pipe into rolled newspaper, and then hit the chimpanzee with the pipe to elicit the desired expression. Eventually, only the presence of the rolled newspaper is needed; the chimpanzee doesn't know that the pipe isn't there, or that they aren't going to be struck. They see it and they wince with a fear grimace. Dogs, too, in fearful situations, will make a fear grimace. They'll wonder: am I going to be hit? Is this a safe situation for me?

Misconstruing aggression, assertiveness, fear, uncertainty, and dominance are easy to do without understanding what certain behaviors mean when paired with other behaviors. So when does a dog show aggression? People talk about "leash aggression," where dogs will bark, growl, or snarl at another dog with teeth bared as they lunge toward another dog. But is this aggression, or is this insecurity? If those two dogs met off-leash, would the interaction be different? Dogs, like chimpanzees, often feel protective over their home territories. Chimpanzees will patrol the boundaries of their territory, and some dogs, too, will be vigilant when a stranger approaches the edge of their fence, crosses the perimeter into their yard, or knocks on their door. What kind of aggression is this, when a dog exhibits typical "aggressive" behaviors?

Take an interaction between two dogs at a dog park, for example. A Husky mix, tail held high and wagging, with a relaxed jaw and forward-facing ears, rapidly approaches a Beagle. The Beagle's ears are back, he's low to the ground, and as the Husky draws nearer, he pulls his lips back in a snarl. The Husky's owner calls his dog back, believing that the Beagle is an aggressive animal. But it's far more likely that the Beagle, seeing the larger individual approaching at a rapid clip, has either misread the Husky's friendly approach signals and is uncertain of the interaction with him, or perhaps this is his first trip to a dog park and he's unfamiliar with and unsure of this environment. Take another interaction at this same dog park: one of the dog owners throws a ball to the Husky, who chases after it, but then a German Shepherd mix swoops in and gets the ball first. The Husky approaches that dog, who is laying on the substrate with the ball between his paws as he chews on it. As the Husky draws nearer, the shepherd begins to growl with exposed teeth. The Husky's owner again calls his dog back, believing that the shepherd is being aggressive, and in this instance, the owner is correct. In the second instance, though, there's a monopolizable resource (only one ball) that two individuals want; one of the individuals shows aggression in order to attain this resource. Is the German Shepherd dominant? In this social instance, he is – he has possession of the highly valued item, while the Husky doesn't. But that doesn't necessarily mean that the shepherd is a "dominant" individual, in relation to the Husky; to determine this, one would have to observe multiple interactions between the two in a variety of behavioral contexts.

When do humans who typically aren't "aggressive" exhibit "aggressive" behavior? Consider, for example, road rage. A person might feel safe within their car, and people will tend to act far differently than they would if they were walking on a sidewalk. Imagine that same level of "rage" if someone incidentally cut you off on the sidewalk vs. if they cut you off while driving. It's a far different context. Similarly, people online will act more aggressively when they have a veil of anonymity than they would during face to face interactions. Dogs, too, will act aggressively in certain contexts but not in others. Take leash aggression, for example: two dogs meeting while restrained by leashes are likely going to act far differently than if they met on their own terms, off-leash, and in a neutral location. So, too, will dogs act differently if one of them is leashed and the other is unleashed; the social context is different, and thus the behavior is different. Leash aggression and leash reactivity, which include a suite of behaviors such as lunging, barking, and growling, result from dogs feeling uncomfortable and out of control during their on-leash social interactions.

Jack is far more likely to aggress toward another dog while he is on the leash and they approach him while off-leash (and often at a rapid clip) than when he is off-leash. Despite many other dog owners' distant (because their dog has raced off without them) assurances that their dog is "friendly," this isn't how most dogs would like to meet another dog for the first time. One dog (the charging, off-leash one) has the upper paw in the situation, while the other dog is restrained and perhaps also feeling protective of their owner. Is the new dog charging at them in excitement? Aggression? Play? Jack is also more likely to be unfriendly toward another dog while they are both on-leash, but in situations where both dogs are off-leash, his behavior (assuming a dog who's initiating and reciprocating friendly behaviors) is friendly.

Aggression can be situational, but is it a trait that's heritable? According to recent research, aggression is the result of both environmental and genetic factors, although the literature is still lacking on the latter. Let's revisit the topic of pleiotropy – when you inherit one trait, you often inherit others, as well. Animals that have been bred for a particular look might also have particular behavioral traits that co-occur with those phenotypic characteristics, and vice versa (as in the case of Belyaev and Sorokina's foxes, which came to resemble domesticated dogs in both their physical appearance and in their behaviors). A dog that is bred to have a certain appearance might inherit genes to be more aggressive, as well.

Environmental factors and individual differences work within a large population of genes, exerting effects that result in different traits. In instances of artificial selection, where individuals have been intentionally bred with one another to exaggerate (or diminish) one trait over another, the heritability of certain characteristics can be increased, resulting in higher rates of genetic disorders, including mental and behavioral syndromes. Specific genetically related aggression issues include Springer Rage, wherein affected individuals show the sudden onset of aggression.

OF MICE AND MEN

"Aggression" is often viewed in a negative light, but aggressive behavior occurs in a wide range of species, including insects, fish, amphibians, avians, and mammals. The prevalence of this behavior throughout the animal kingdom indicates likely fitness benefits and heritability of aggressive tendencies. Mice and humans are separated by 75 million years or more of evolutionary history, but because they share homologous genes, mice remain the go-to model for the comparative study of genetics, including the heritability of aggression. Common rodent models for aggression studies have included the mouse (*Mus musculus*), the rat (*Rattus norvegicus*), the prairie vole (*Microtus ochrogaster*), and the hamster (*Mesocricetus auratus*). Through selective breeding, it is possible to select for genes that result in individuals that exhibit more aggressive behaviors than those who aren't bred specifically for this temperament, demonstrating that aggression has biological, environmental, and temporal factors. A 1983 study with mice resulted in animals that were highly aggressive, but the emergence of this aggression was dependent upon specific developmental periods. The mice that had been bred to be aggressive later attacked other mice, apparently without provocation, and at far higher rates than the mice who had not been bred to be aggressive.[141] Because mice and humans share homologous genes, these studies have also helped identify the role of neurotransmitters in human aggression (Fig. 40).

Beyond mice, studies have now started to examine the genetics of canine aggression, as well.[142] Cases of canine aggression can be particularly problematic for dogs

FIG. 40

Mouse. *Sketch by Emilia Galletto.*

[141] Cairns, R. B., MacCombie, D. J., & Hood, K. E. (1983). A developmental-genetic analysis of aggressive behavior in Mice: I. Behavioral outcomes. *Journal of Comparative Psychology, 97*(1), 69–89.
[142] Duffy, D. L., Hsu, Y., & Serpell, J. A. (2008). Breed differences in canine aggression. *Applied Animal Behavior Science, 114*, 441–460.

and their owners; aggressive behaviors that go unabated and cannot be redirected often result in the dog being relinquished or euthanized. Using the Canine Behavioral Assessment and Research Questionnaire (CBARQ), the owners of 30-plus dog breeds answered questions about how their dogs responded to different situations and stimuli. The CBARQ, which was created by the researchers with the Center For Interaction of Animals and Society at the University of Pennsylvania, was designed to provide standardized evaluations of canine temperament. There were eight breeds that demonstrated similar levels of stranger-directed aggression. These breeds included Golden and Labrador Retrievers, Dachshunds, English Springer Spaniels, Poodles, Rottweilers, Shetland Sheepdogs, and Siberian Huskies. Some of the breeds in the study, such as Dachshunds and Chihuahuas, had higher than average human-directed aggression, while some breeds, such as Akitas and Pit Bull Terriers, demonstrated higher levels of dog-directed aggression. Aggressive behaviors appeared to be most severe when they were dog-directed or directed toward unfamiliar people. The dog breeds that exhibited the most severe forms of human-directed aggression, including biting, included Jack Russell Terriers, Chihuahuas, Dachshunds, Australian Cattle Dogs, Beagles, and American Cocker Spaniels. The dog breeds that exhibited the most severe forms of dog-directed aggression included Jack Russell Terriers, Akitas, and Pit Bull Terriers. The dog breeds that exhibited the lowest rates of aggressive behaviors were Labrador and Golden Retrievers, Greyhounds, Whippets, Bernese Mountain Dogs, and Brittany Spaniels.

BREED-SPECIFIC LEGISLATION

What do Dobermans, Rottweilers, and Pit Bulls all have in common? At one time or another, each of these breeds (and types) of dogs has been on the receiving end of discrimination via breed-specific legislation (BSL). But BSL is legislating the wrong end of the leash and punishing dogs based upon their appearance and not on their temperaments or behaviors. Pit bulls are the current posterchild of BSL. They're often targeted in the media, as well – how often does a negative story about a dog lead with "Pit bull bites child?" If it was a Labrador, or a Poodle, or a Pomeranian, the breed isn't just left out of the title; it's usually omitted from the article altogether. Pit bull type dogs (along with a slew of other breeds), are considered by insurance companies to be "aggressive," but research shows that aggression can be demonstrated by a variety of different dog breeds and in a variety of contexts (dog-directed, human-directed, stranger-directed, etc.). "Pit bull" is also used as an umbrella term comprising dogs that are genetically considered to be pit bulls, such as American Pit Bull Terriers, American Staffordshire Terriers, American Bullies, and Staffordshire Bull Terriers. "Pit Bull" also sometimes includes the American Bulldog, Miniature Bull Terrier, and Bull Terrier. In addition to including dogs that are genetically "pit bulls," however, this category also includes dogs that have "pit bull like" appearances, including square looking heads and stocky bodies. This creates issues with overrepresentation of "pit bulls," as both phenotypic and genetic criteria are being used. In other words, those who are categorizing "pit bulls" are double dipping into the categorizing. It would be like saying someone could be Irish if they had Irish genetics or if they had red hair, freckles, fair skin, or light-colored eyes, when in fact, this could mean that they were instead Scottish, Scandinavian, British, or some combination of Northern European ancestry. If it sounds ridiculous, that's because it is, but many dogs are judged by their appearance every year, subjected to breed-specific legislation that bans them from certain cities.

A growing number of organizations, including the American Veterinary Medical Association (AVMA), the Centers for Disease Control (CDC), the American Society for the Prevention of Cruelty to Animals (ASPCA), and the Humane Society of the United States (HSUS) are speaking out against BSL. Any dog, regardless of their breed or appearance, has the capability of biting and BSL policies do not diminish the number of dog bites. While many jurisdictions are overturning their BSL policies, many discriminatory policies remain entrenched in the legal system, bolstered by dog bite statistics. But these statistics are particularly problematic, as they don't take into account the behavioral context. According to the Center for Disease Control (CDC), 4.5 million people are bitten by dogs annually; of these, approximately 800,000 of them need to receive medical treatment and 20 to 30 of these bites result in death. It's important to know why these bites are occurring, though, and to be proactive in situations where biting is likely to occur (such as with a dog who is anxious and has no "escape," or a dog who is feeling aggressive and frustrated because he or she has been chained up in a backyard 24/7). The United States currently has no standardized dog bite reporting system.

BSL discriminates against dogs who appear to belong to a certain breed without taking into account their actual genetic heritage, their temperament, or their history of past behavior. Rather than creating over-reaching policies, legislators would better serve their communities by enforcing laws about irresponsible pet ownership (such as owners that allow their dogs to roam freely without supervision, or keeping their dogs chained in a backyard 24/7, which increases frustration, aggression, and stress levels for dogs, thus increasing the chance of potentially "dangerous" behaviors). Creating stronger prohibitions against dog fighting, providing incentives for spaying and neutering all animals, and participating in educational outreach about pet behavior, responsible pet care, and being better stewards for our animals are all more effective bite prevention strategies than BSL.

DEBUNKING DOMINANCE THEORY

Powdered wigs. Crinolines. Bellbottoms. Shoulder pads. Rococo. Art Nouveau. Mid-Century Modern. Anthropocentrism. Dualism. Materialism. Whether it's fashion, architecture, or ideology, trends come and go, but the ideological ones are often the most influential – and potentially pernicious. Over the centuries, our beliefs about animals have experienced vast paradigm shifts. Descartes headed the dualist movement, indicating that animals were unfeeling automata. Later researchers, including Darwin, found substantial evidence that supported the existence of emotional experiences among non-human animals. There have been just as many trends and shifts in the field of animal behavior. For decades, "dominance theory" was heralded as the go-to paradigm for dog trainers. Popularized by programs such as "The Dog Whisperer," dominance theory was based upon some flawed studies on wolves. While the misconceptions were later rectified, the damage was already done. "Dominance theory" has done to dog training what the "vaccines cause autism" controversy has done to modern pediatric medicine. In 1998, Dr. Andrew Wakefield published a research paper stating that the measles, mumps, and rubella (MMR) vaccine caused autism and bowel disease. While the results of Wakefield's study couldn't be replicated, and no evidence was ever found to support his claims, investigators did find that Wakefield had a vested interest in a competitor vaccine. The study – and Wakefield – were completely discredited, with Wakefield losing his medical license, but decades later, parents still believe that there's an associated risk between vaccines and autism. This misplaced belief has led to numerous outbreaks of diseases, such as measles, that could be completely eliminated with vaccines.

During the 1930s and 1940s, animal behaviorist Rudolph Schenkel studied captive wolves in the Basel, Switzerland Zoo, later publishing his work, "Expressions Studies on Wolves," in 1947. His studies, which used unrelated wolves, appeared to show that wolves exhibited a social dominance/alpha model in their packs, based upon their interactions with one another. The studies were flawed not only for their relatively short time frame, but also on their over-reliance on the behavior of unrelated individuals in an artificially created social situation. This was more "reality television" than family documentary – this wasn't how a normal wolf family would relate to one another. Given the artificiality of the situation, certain behaviors – including agonistic ones – were also emphasized over others, providing the foundation for the "dominance theory model."

This concept of wolves having an "alpha" who led the group was reinforced in 1970 by wildlife biologist David Mech's book, *The Wolf: The Ecology and Behavior of an Endangered Species*. Mech's tome on lupine behavior expanded upon Schenkel's competition-based pack hierarchies, wherein he stated that wolves dominated one another. From 1958 to 1962, Mech, then a graduate student at Purdue University, studied wolves on Lake Superior's Isle Royale, which has a unique single predator (wolves) single prey (moose) relationship. This large prey source was defendable and required large social groups to bring them down. This created the perfect environment for intense inter- and intra-group fighting; not only was there one primary prey species, but the availability of prey also varied widely. The island has seen dramatic fluctuations in the populations of both the wolves and the moose; in 1980, the wolf population plummeted after humans accidentally introduced canine parvovirus, and 16 years later, the moose population crashed after the dual strikes of the harshest winter on record and an unprecedented moose tick outbreak. Isle Royale has a highly variable kill rate; in 1992, it was twice the average kill rate, but in 1981, it was one-third the typical rate. There was a corresponding wolf population spike in 1992 and a dramatic decline (from 30 to 14) in 1981. Thus, with a lone, large prey species that's wildly inconsistent in its availability, it's the perfect environment for increased fighting and "dominance" behaviors.

Much to Mech's chagrin, he later learned that the wolves that he studied were behaviorally unique due to their unusual geographic location and the resulting predator-prey relationships. Mech recanted the hypothesis that they had an "alpha," but instead had a dominance hierarchy headed by the father of the family group. Mech stated that the "alpha" concept was in fact an artifact of captivity. In 1999, Mech wrote: "I conclude that the typical wolf pack is a family, with the adult parents guiding the activities of the group in a division-of-labor system in which the female predominates primarily in such activities as pup care and defense and the male primarily during foraging and food-provisioning and the travels associated with them."[143]

Unfortunately, the dominance damage had been done long before the 1999 paper clarifying the sociality of wolves came out. The late animal behaviorist Dr. Sophia Yin wrote that "dominance" in animal behavior is a relationship between individuals that was established by means of force, aggression, and submission, wherein the dominant individual(s) have preferred access to limited, highly desirable resources, such as mating opportunities, food, and preferred areas. In evolutionary terms, hierarchies can be important because the more dominant individuals have greater opportunities to pass their genes on to the next generation than do more submissive individuals. Interestingly, Dr. Yin notes, the ancestors of modern domesticated dogs lived as scavengers with a promiscuous mating system for the past 10 to 15,000 years, while wolves lived within a social system headed by mother and a father and their offspring and other close relatives. In the latter social system, only the mother and father wolf (the highest-ranking members of the pack) would have mating opportunities.

Dominance hierarchies have been studied in a number of species, including lions, macaques, chimpanzees, meerkats, wolves, and chickens, the last of which inspired the phrase "the pecking order." Dominance is not a separate personality trait; if three individuals who were dominant in their own social groups were then placed together, they will struggle with one another to determine who is the first, second, and least dominant member of the group. This is likely what occurred during the 1940s wolf studies that disproportionately influenced the "dominance paradigm" as applied to domesticated dogs. But different species have different hierarchies and can be either egalitarian or despotic. In species that have an egalitarian, linear ranking system, each individual knows his or her status in relation to every other individual in the group. In species that have a despotic system, one individual is dominant, while all others are equally submissive. Species that exhibit a despotic social system include gorillas, meerkats, and gray wolves – but not domesticated dogs. Gray wolves don't earn their "alpha" status from fighting, like many other despotic species do; the "alpha male" is typically the father of the family and his offspring naturally defer to his leadership.

---

[143] Mech, L. D. (1999). Alpha status, dominance, and division of labor in wolf packs. *Canadian Journal of Zoology*, 77(8), 1196–1203.

Similarly, dog owners need to lead their dogs not like despots, but more like dads. While "dominance theory" was trendy among dog owners and trainers for decades, recent research discredits its widespread use for domesticated dogs. Dogs present formal dominance, which is demonstrated by context-independent status signals; as a result, formal dominance and submission do play important roles with dog-dog relationships and possibly with humans, as well.[144] When humans enforce a "dominant" status with their dogs, however, this can be problematic, and quantitative data in support of using the dominance theory for domesticated dogs is limited.[145]

Despite the prior popularity of using dominance theory to train a dog, there are many adverse effects with this. Dominance Theory is the cornerstone of the "Dog Whisperer," Cesar Millan's, work, and he refers to it throughout his training sessions. Millan's philosophy centers around pack leadership, with the human as the "alpha dog" – it's dominance theory in its purest form, and it's purely misused. Millan frequently uses aggressive techniques when training dogs, including choke collars, strikes, choking off dogs' oxygen by pulling up their collars, and "alpha rolls" to subdue dogs and show them who's the alpha. A 2006 *New York Times* article stated that Millan's "outdated" methods "…might make good television…but flies in the face of what professional animal behaviorists…have learned."[146] In April of 2010, the PBS documentary series *Through a Dog's Eyes* aired a segment entitled "The Dominance Myth." It stated, in part, "Scientifically, dominance makes no sense."[147] Similarly, best-selling author and dog behavior expert Dr. Patricia McConnell notes that confrontational techniques elicit aggression, not respect, from our dogs. McConnell noted that using confrontational techniques elicits the highest levels of aggressive responses from dogs, with 43% of dogs responding aggressively after being hit, 38% to having an owner grab their mouth or forcefully remove an object, 36% to putting on a muzzle, 29% to an alpha roll or "dominance down," and 26% to a jowl or scruff shake.[148]

When people misunderstand what "dominance" means, they will attribute certain behaviors to the dog "wanting to be dominant" rather than the dog just being a dog. For example, there's a common misconception that your dog shouldn't go through a doorway before you. But why? Where did this belief come from, and why do people believe that this makes a dog "dominant?" It stems from a basic misunderstanding of behavior in general and what dominance is and is not. If your dog goes through a door before you, he or she is just excited to get to what's on the other side of that

---

[144] Schilder, M. B. H., Vinke, C., & van der Borg, J. A. M. (2014). Dominance in domestic dogs revisited: Useful habit and useful construct? *Journal of Veterinary Behavior, 9*, 184–191.

[145] van der Borg, J. A. M., Schilder, M. B. H., Vinke, C. M., & de Vries, H. (2015). Dominance indomestic dogs: A quantitative analysis of its behavioural measures. *PLoS ONE*, https://doi.org/10.1371/journal.pone.0133978.

[146] www.nytimes.com/2006/08/31/opinion/31derr.html.

[147] www.pbs.org/show/through-a-dogs-eyes.

[148] www.patriciamcconnell.com/theotherendoftheleash/confrontational-techniques-elicit-aggression.

doorway – nothing more, nothing less. They aren't thinking, "I'm the 'alpha,' so I get to go through the door first." It's probably something more along the lines of, "Oh, is this where we're going now? Okay, you're going there, so I'm happy to go there, too." Jack almost invariably goes through doorways first, but it's because of his gait (it's hard to walk slowly on three legs) and completely unrelated to dominance. Jack often waits after crossing the threshold to see if the humans accompanying him are still following.

Other strange – but popularly entrenched – beliefs include not allowing your dog to eat before you do (because the "alpha" should always eat first), not allowing your dog to hold eye contact with you, not being the one to initiate a greeting (your dog should greet you first, because you are the "alpha"), refraining from interaction with your dog unless you are actively training him or her, and refraining from showing signs of affection with your dog. Having learned thus far in this book how dogs perceive the world, how they have co-evolved with humans for millennia, and how tuned in they are to their human companions, dogs must find rules such as these confusing at best and at worst, the root cause of anxiety and depression.

On the practical, clinical side, outdated, inappropriate "dominance" concepts in the hands of owners and trainers has created a significant percentage of the behavior problems and issues seen by specialists. Fear and anxiety behavior is the common outcome and it's likely the cause of the current epidemic of fear and anxiety aggression being seen in the dog population in the United States. This results from a direct combination of serious inbreeding issues in many breeds and the inappropriate use of "dominance" methods.

Now, to be clear, lumped together in this concept of dominance, as a result of Cesar Millan and his followers, are outdated "pack theory" concepts (such as "You need to be a better leader") and inappropriate methods for expressing outdated "pack theory" ideas: "dominance signals." These so–called expressions of dominance ("alpha-rolling," pinch and prong collars, and "helicoptering") are in reality simply forms of punishment.

So, when JCH sees a behavior case, one of the many important pieces of information is a history of learning exposure: how has the owner attempted to teach this dog, and did the methods involve some combination of pack theory methods, punishment (whether disguised as "dominance signals" or outright aversive methods), or positive reinforcement, in its many incarnations.

As we have discussed, pack theory and hierarchies are a method for reducing conflict over limited resources, providing an orderly, less-dangerous method for resolving conflicts. It's not a training or general management method, and no dog understands it to be this way. They are useless in the sense that they are used by that community. Do dogs need structure? Do dogs need rules and limits? Do dogs need to be taught the ins and outs of living with humans, and other dogs? Of course, and to this extent, the owner needs to be a leader, but not necessarily through the use of aversive methods.

The second option is punishment training, or more appropriately, the use of aversives. The proper use of aversive training in subjects well-chosen to be appropriate for this type of training (of a certain temperament type!) and in the hands of a skilled aversives trainer is highly effective. Punishment is a strong quadrant of the four fundamental types of operant learning. But in the hands of the inexperienced trainer or owner, and used on a dog with an inappropriate temperament, aversives are extremely damaging. JCH sees many, many cases in which inappropriate application of aversives (people will say, "But I saw it on TV!") is at least involved as a contributing factor. Inappropriate application of aversives commonly produces fear and anxiety, and anxiety-based aggression is one of the most common behavioral issues in dogs in recent years.

Treatment for these issues generally involves the education of owners and trainers to stop the use of the aversive methods, and then either a slow decline in symptoms as the dog gradually learns new contingencies ("When I do this or that, I no longer get punished"), or a more rapid improvement through the establishment of positive-reward-based connections ("When I do this or that, something positive happens!"). Basically, the dog has learned to fear the owner, or trainer, and will no longer perform the behavior in her presence ("Look, it worked!") but for the wrong reasons and having suffered damage to their future relationship. The aversive result is not simply focused on the recently-performed unwanted behavior, as desired, but effects the social relationship more broadly. Dogs prone to do this would be weeded out of a well-run organization using aversive methods.

What are the implications for the use of pack theory, and its associated inappropriate dominance signals, or inappropriate application of outright aversive training methods? We see a lot of anxiety and fear-based behaviors (aggression, but also house-soiling) result from these methods, but an even more dangerous, and common, outcome is redirected behavior, especially aggression. The dog becomes fearful of the handler (inappropriately) using aversive methods, and no longer performs the undesired behavior in their presence (success!) but will redirect that anxiety and fear towards a less fear-inducing target, often women, children, or other pets. JCH sees this not uncommonly in his clinical experience: An owner is using punishment methods, and the dog behaves well for them, but "mysteriously" starts to growl at the kids, or the other dog, or the cat. The dog is redirecting its frustration, its anxiety at a "lesser" target, and this redirected aggression can become very dangerous if not brought under control quickly. I have seen numerous cases where the dog begins to redirect at the wife, punishment is used to stop that, and the dog just works its way down the list of less intimidating targets, usually until a bite occurs.

Most conflicts between dogs and their people derive not from dogs' desire to be dominant, but from fear or reactivity or their desire for consistency, reassurance, and guidance. It's important to differentiate between "aggressive" behaviors and "dominant" individuals and not conflate the two terms. As demonstrated with the mice and dog genetic studies, aggression can have a genetic component, but temporal, hormonal, and situational elements can determine the occurrence of an aggressive event, as well. Female hyenas are considered to be aggressive, compared to male hyenas and to other mammals. Not only do they have a lot of testosterone, relative to other female mammals, but they have male-looking genitalia, as well. Aggression describes the behaviors of an individual, while dominance describes the relationship between two or more individuals. An individual who is dominant in a certain social situation may or may not act "aggressively."

In the use of positive reinforcement methods, I (JCH) always say that the worst that you can get from inappropriate application is a fat, poorly trained dog. But not a fearful, anxious dog redirecting its aggression towards family members. In my mind, the potential consequences of inappropriate application of aversive-based methods (regardless of the rationale behind them) dictates that it is irresponsible to recommend them to owners or unskilled trainers.

An organism's environment, including the distribution of resources, plays a strong role in their social structure and behavior. With tens of thousands of years of co-evolution, it should come as no surprise that dogs are behaviorally different from wolves and from all other animals, wild and domesticated alike. While dogs do understand dominance, its misuse often leads to the degradation of the human-canine relationship.

THE RESOURCE HOLDING MODEL

How do we quantify how much an animal can win or protect if he or she were in a battle to the death? The resource holding model, also referred to as resource holding potential (RHP), refers to how much an individual might lose if he or she were in a fight. In 1974, British evolutionary biologist Geoff Parker created this model to disambiguate one's ability to physically defend a resource from one's motivation to persevere in this altercation. Parker's work focusing on behavioral ecology was pivotal in the paradigm shift from viewing evolution primarily in "species survival" terms to more gene-centric ones, with a focus on individuals and their relatives (kin selection).

The "hawk-dove" game, also called "chicken," is one of the models in game theory. It is based on the principle that both players will benefit if one of them yields, but the players' choices are dependent upon each other. If Player A yields, Player B should not, but if Player A doesn't, Player B should. For example, if two cars are driving directly at one another, both drivers are going to "lose" if neither yields and they crash into each other. "Hawk-Dove" refers to the competition for a resource. The two players can either be conciliative or conflictive. RHP is an animal's ability to win in an all-out fight and is often referred to as a prime example of canine social structure, as canine social behavior is fluid.

CHAPTER 10

The tale of our first friends: Using the natural history of domesticated dogs to understand their behavior

BIG FAT ROTTEN COWS, DRAGONS, AND PILTDOWN MAN

Nine-year-old Jude Sparks was descending from a sloping bluff in the Las Cruces, New Mexico desert in 2016 when his foot struck something unusual. He bent over to examine the large, unfamiliar object protruding from the earth. Mystified, he showed his brother, Hunter, who assured him that the find was only a "big fat rotten cow." This wasn't an ill-fated bovine, however: it was the fossil of a 1.2-million-year-old animal called stegomastodon, a distant cousin of the woolly mammoth and the elephant. Jude's rare find quickly found its place in a local museum. In the 4th century BCE, Chinese historian Chang Qu considered his own version of the "big fat rotten cow" when he examined a giant fossil discovered in what is now modern-day Sichuan Province. But Qu didn't believe that this was a cow; he was convinced that the 30-foot-long stegosaurus, complete with spikes and armored plates, was a "dragon." Finds like these likely fueled the ancient belief that dragons had once roamed the world. And in 1912, amateur archaeologist Charles Dawson reported that he had unearthed a fossil in Piltdown, Sussex, that was the "missing link" between humans and apes.[149] The mandible, which looked more like that of a modern nonhuman ape, was paired with a skull that appeared to be from a prehistoric human. The fossils were placed in a museum for public viewing, but later analysis found that the bones were not that of a nonexistent "missing link," but the combination of a modern Orangutan jaw bone and a medieval human's skull. Dawson had apparently tried to "create" a fossil that would fit his own narrative of evolution, and it wasn't the first time that he had fraudulently found an artifact. Piltdown Man (*Eoanthropus dawsoni*), was his magnum opus, a hoax that lasted 41 years, long after Dawson died in 1916 from septicemia. He was posthumously implicated in the scheme.

---

[149] Humans are apes, thus Dawson not only poorly perpetuated a hoax, but he failed to properly address his motley specimen.

Hoaxing is nothing new in the world of science, though – John James Audubon pranked fellow naturalist Constantine Samuel Rafinesque when he sketched and described 11 nonexistent fish species. Among Audubon's counterfeit creations were the "Devil-Jack Diamond fish," a creature that was four to ten feet long and sported bulletproof scales (Fig. 41). Audubon's hoax was so effective that Rafinesque included the unlikely animal in his book, *Icthyologia Ohiensis, or Natural History of the Fishes Inhabiting the River Ohio and Its Tributary Streams (American Environmental Studies)*. Rafinesque, in turn, fabricated the Walam Olum, or "red record," which was purportedly a "Native American Creation Epic" that he scribed in 1836 in his book, *The American Nations*. The Walam Olum includes an origin of the universe story and also suggests that there was a mass emigration 3,600 years ago across the Bering Strait. Rafinesque claimed to have translated this story from pictographs on birch bark plaques which were "lost" shortly after the translation occurred. Later researchers were unable to find any evidence to substantiate his claims. Like Dawson, Rafinesque's historical hoax didn't come to light until well after his death.

Humans have a long legacy of scientific mistakes and missteps. From misunderstanding our evolutionary place in the world – and our spot in the animal kingdom – to phrenology and from maps of a flat earth to believing that nonhuman animals were devoid of emotion, our errors have been charted in our museums throughout the generations. Even as we have sought to better understand our world, we have often obfuscated the truth. Throughout history, humans have discovered unfamiliar plants,

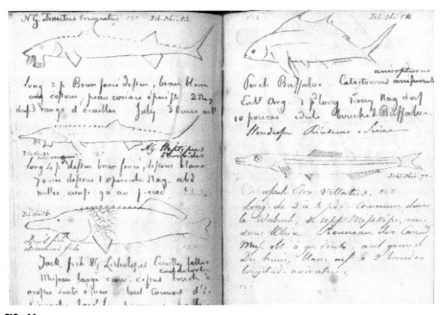

FIG. 41

Devil-Jack Diamond fish.

animals, and bones, and we have tried to classify them – or we have tried to create narratives to fit into our own view of the natural world. The earliest archaeologists, like Qu, happened upon huge fossils that didn't correspond to any of the extant fauna that they had ever seen. Many of the earliest detailed records of the living beings that we share the planet with were nothing more than wild guesses. Greeks and Romans believed that elephants fought with dragons, Aristotle was convinced that eels spontaneously generated from the mud, ancient civilizations thought that crocodilians were merely scaled monkeys, and seventh century writer Isidore of Seville claimed that pelican blood could resurrect the dead. From maps that depicted nonexistent sea monsters to phrenological maps of the human head, mankind has had many scientific missteps along the way, but while wildly inaccurate, these humorous attempts at natural history documentation revealed man's fascination with the world and a strong desire to understand it.

HISTORIA NATURALIS

Housed within the venerable walls of the Smithsonian Institute of Natural History in Washington, D.C., you'll find artifacts from a wide range of fields that reflect the intellectual curiosity of modern humans. From anthropology to vertebrate zoology, from botany to paleobiology, and from entomology to mineral sciences, there are now more than 145 million artifacts in the Smithsonian's collections. The Smithsonian houses everything from Moon rocks to the Hope Diamond, from the body of World War I carrier pigeon hero Cher Ami,[150] to *T. Rex* fossils, from Indiana Jones' fugitive Fedora to Dorothy's ruby red slippers, and from Teddy Roosevelt's teddy bear to ancient dog collars. The Smithsonian's Natural History collection includes many artifacts depicting our storied relationship with dogs, including skeletons, collars, harnesses, carvings, paintings, and even taxidermied dogs such as Owney the United States Postal mascot, who lived from 1887 to 1897, and Sergeant Stubby, the official mascot of the 102nd Infantry Regiment of the U.S., who lived from 1916 to 1926. This stunning array of artifacts chronicles the natural history of our world (Fig. 42).

Natural history is the multi-disciplinary study of organisms within their environment, comprising their evolutionary history, their role in the environment, and their behavior, including social relationships. The natural history of a species elucidates their behavior, including their social structure, mating choices, and sociality. Throughout history, adventurers, naturalists, and biologists have recorded and described the flora and fauna of the world. Commissioned by President Thomas Jefferson, Meriwether Lewis and William Clark set off in 1804 to document the unexplored new territories of the United States. Their two-year scientific sojourn yielded hundreds of botanical and zoological specimens: During the Lewis and Clark Expedition, which spanned May 1804 to September 1806, Lewis described

---

[150] Cher Ami was said to save the lives of 200 soldiers.

FIG. 42

The Smithsonian Institution.

122 previously undocumented animal species, including grizzly bears, mountain beavers, and coyotes; he additionally described 178 previously undocumented plant species. Among the canines documented during the expedition were the Coyote (*Canis latrans*); Gray wolf subspecies, including the plains form, (*Canis lupus nubilus*), Gray wolf (*Canis lupus irremotus*), Gray wolf (*Canis lupus fuscus*), Swift fox (*Vulpes velox*), and the Red fox (*Vulpes fulva macroura*). Lewis and Clark were also accompanied on their journey by a Newfoundland named Seaman, who was credited with "saving" their mission when he alerted the slumbering humans to a rogue buffalo bull who was charging through their camp at night. Lewis and Clark followed in the long tradition of scientists and philosophers who had observed and

documented the natural world, from Aristotle to Linnaeus and from Georges-Louis Leclerc, Comte de Buffon, to Alexander Humboldt, whose abundant research influenced later naturalists, including John Muir, Henry David Thoreau, and Charles Darwin.

During his time on the *Beagle* from 1831 to 1836, Charles Darwin was a documentarian of all things wild, wonderful, and as-yet-unfamiliar to the scientific community, from Falmouth to the Falkland Islands and from the Galapagos to the Cape of Good Hope. As Lewis and Clark did a quarter of a century before, Darwin discovered and documented new animal species, including numerous mockingbirds and finches. But in his epic tome, *The Origin of Species*, Darwin appeared to be fascinated not only with his new discoveries, but also with the natural history of a previously familiar species: the domesticated dog. Over his lifetime, Darwin lived with more than one dozen dogs of his own, and they became a rich resource for his studies on evolution, artificial selection, and emotion.

Several hundred years ago, there were probably more than five million wolves on the planet; today, there are as few as 150,000 left. Domesticated dogs, however, are globally prolific; today, there are hundreds of millions of domesticated dogs worldwide. Dogs joined humans prior to the advent of agriculture and have been culturally and socially important to humans even before they first entered our homes. For ancient Mesopotamians, dogs symbolized Ninisina, the goddess of healing and medicine. For Neo-Babylonians, dogs were emblems of magical protection. Many cultures believe that dogs are said to protect us from dangers, both above and below. Throughout mythology, dogs have guarded the gates of the underworld and Sirius, the dog star, is the brightest star that's visible from Earth.

Throughout their history with us, dogs have been depicted in our art, literature, and lore. Some of the earliest known cave art depicts canines. Wolves graced the 17,000-year-old cave paintings in Font-de-Gaume, France. A series of rock carvings from the Arabian Peninsula depicts a familiar scene that could've occurred during recent times: a hunter, armed with bow and arrows, is flanked by several dogs, some of them on leashes. The image depicts other hunters, also accompanied by dogs. It could be modern…except it dates back more than 8,000 years ago. The dogs, who were carved into rock faces in Jubbah and Shuwaymis in what is now present-day northwestern Saudi Arabia, look much like modern dogs do: they have medium builds, curled tails, and short muzzles, and closely resemble the Canaan breed. Cave paintings from the Tennessee Cumberland Plateau dating back 8,000 years depict numerous canid species, including animals that appear to be jackals, foxes, and wolves. And while the famous paintings in the Chauvet-Pont-d'Arc Cave in the Ardeche region of Southern France don't depict canids, they do include those famous footprints of a human child and a large canid companion. The prints, which are approximately 26,000 years old, are considered to be some of the oldest known physical evidence of the human-canine relationship. Interestingly, these canid prints already hint at the effects of domestication: the front paw pads have a shortened middle digit, a trait shared by domesticated dogs, but not wild canids.

THE LOST BOYS

The foliage parted and a child emerged from the edge of the forest on all fours, his hair disheveled, his skin stained with dirt. He was flanked on each side by two wolves, who paused for a moment and sniffed the air with their noses. The child, poised on bent arms, did the same, and then scanned the horizon. He looked more canine than human, and perhaps, by this point, he was.

The ancestors of modern dogs weren't the only ones who chose to live in the home of another species. This story of a feral boy isn't an apocryphal tale; incidents just like these, where children somehow became lost in the woods and were later raised by wild animals, including canids, have been documented throughout history. In 1867, Dina Sanichar was discovered in a cave in the Bulandshahr District of India. A group of hunters believed that the prone form in the cave was a wild animal, but discovered that it was instead a six-year-old boy who had been living with a pack of wolves. Despite the efforts of missionaries, Sanichar, who came to be known as the "Wolf Boy," never learned to speak. He still preferred to follow the behaviors of the wolves who had been his family. Sanichar died in 1895 and it's possible that he was the inspiration for Rudyard Kipling's "Mowgli" in *The Jungle Book*. In 1946, Marcos Rodriguez Pantoja, then 19, was discovered in Southern Spain living with wolves. Pantoja had been sold to a goatherder as a small child, and after the man's death, Pantoja fled to the Sierra Morena, where he lived with the wolves for one dozen years before he was discovered by the outside world. Pantoja returned to living with humans at 19, but it wasn't an easy transition. In addition to these documented cases, there are also literary legends about feral human children being raised by canids. Rome was said to have been founded by the twins Romulus and Remus, who had been raised by a wolf.

Many of these "wild children" adopted the locomotion, behavior, and communicative patterns of their canine families, sometimes living with the wild canines for years. Reintroducing them to human society was typically a failure, as these feral children were now more dog-like, culturally, than they were like humans. Many of them had missed out on critical and sensitive periods of socialization with human peers, similar to the beach dogs like Gus Garner who had missed out on these early opportunities.

FROM WOLF TO DOG

For millennia, dogs have symbolized faithfulness, fidelity, love, loyalty, guidance, and protection. Today, the global domesticated dog population is now nearing one billion, many of which are feral scavengers in undeveloped countries. In the United States alone, 54.4 million households have at least one dog. The modern dog is the first known domesticated species, and as such, it is a fascinating study in human behavior, evolutionary processes, and comparative behavior. It's also a benchmark of natural history studies. Tens of thousands of years ago, prior to the advent of agriculturalism, the ancestor of modern dogs split off from the lineage of the ancestor of modern gray wolves. Over many generations, this eventually led to the modern domesticated dog. In 1758, Carl Linnaeus released his tenth edition of *Systema Naturae*, which was the first work that consistently applied binomial nomenclature (the formal naming system with a two-part Latin name) to describe the animal kingdom. In this edition, he noted that dogs were a different breed than wolves and classified the domesticated dog as *Canis familiaris*, which translates to "dog-family," or the family dog, while the wolf was *Canis lupus*, or "dog-wolf."

GLANCES WITH WOLVES

To understand the significance of the natural history of the dog, one also has to consider the natural history of their closest relatives, the gray wolf. Gray wolves, like dogs, are highly social creatures, which live their lives in packs consisting of families. The packs, which usually average in size from four to 11 individuals, are led by a breeding male and female (the parent wolves) and typically their offspring. Non-related individuals also join a pack from time to time. Members of a wolf pack participate in all activities: they share pup-rearing responsibilities, hunt, rest, and eat with one another. According to the Washington Department of Fish and Wildlife, wolves in the wild usually live for only five years, a truncated lifespan reflecting high predation levels from humans, both legal and illegal, fatal injuries sustained from vehicle collisions, and starvation, disease, and injuries sustained while hunting prey. In ideal circumstances, wolves have been known to live for 15 years or longer. Most wolf packs produce only one litter per year, thus having a relatively slow population growth, and most litters consist of four to six pups.

Wolf Haven International, located in rural Tenino, Washington, has provided a home for more than 200 captive-born canids, including wolves, wolf-dog hybrids, and coyotes, since 1982. The sanctuary provides educational outreach and advocates for both captive and free-living canids. Wolf Haven's director of animal care, Wendy Spencer, has been with the organization since 1998. She's proud that Wolf Haven has a unique mission of rescue, advocacy, education, and public outreach about wolves' species-specific behaviors. The organization is one of only a handful that doesn't socialize its animals, as many of them will be returning to the wild. Spencer has found that the likenesses (both real and perceived) between wolves and dogs has been problematic for wolves and wolf conservation.

"There are so many similarities between wolves and dogs, but wolves are genetically different from domesticated dogs; they're hard-wired differently," she said. As a result, dog behavior doesn't translate directly to wolf behavior. "Knowing dogs and dog behavior doesn't mean you'll also know wolf behavior. Any behavior you will see in a dog will also be expressed with wolves, but they're often in a different behavioral context and more intense, as well."

The gestural repertoire of the dog doesn't always have a direct equivalent with wolves. Dogs have a variety of information-laden gestures, such as the "play bow," that are exhibited across behavioral contexts. The play bow is an invitation to play – and it's universally expressed among domesticated dogs, much like chimpanzees use a "kickback" gesture to invite their peers to play[151] – but with wolves, you have to look at other indications, as well. It's like a language that has similar words, but emphasizing different syllables changes the entire meaning.

"You have to look at the whole picture," Spencer said. "While a wolf might be in the play bow position, you also have to see what their ears and tails are doing. Are the hackles up? You can't just assume that they're soliciting play."

While dogs have a wider range in their vocal repertoire, there are also similarities in their auditory communication. "With some feral dogs, you'll hear a sneeze or a huffing behavior," Spencer said, "and those are alarms. In close proximity, or when they're unsure, wolves will also 'sneeze' or make a huffing sound. These vocalizations will escalate from 'sneeze,' to huff, and then to an alarm bark. When they're in the wild, wolves have the opportunity to retreat, but when they're in captivity, that's not always an option."

Wolf communication can be subtler to discern than domesticated dog communication, and a lot of information can be conveyed in glances that humans don't notice and in sounds that are inaudible to our ears. Wolves, like chimpanzees, change their behavioral contexts at a more rapid pace than their closest cousins. "Wolves' behavioral changes can seem quick, and like they happened without warning, but there's a nuance to them. We just aren't aware," Spencer said.

While Wolf Haven International is helping with conservation efforts, the territory of gray wolves continues to shrink. The original range of the gray wolf extended across the Northern Hemisphere from the Arctic circle through Southern Mexico, southern Asia, and northern Africa. Modern day gray wolves currently inhabit a few isolated areas of the contiguous United States, Alaska, and Canada, with very small populations remaining in Mexico and Eurasia. Wolves typically range from 140 to 400 square miles and disperse about 60 miles from their natal packs, but have been known to travel as far as 500 miles. Wolves live in a variety of habitats, including tundra, forest, scrub forest, taiga, and mountains. They hunt and scavenge on carrion and will opportunistically eat domestic livestock, creating friction between conservationists and cattlemen who use public lands to graze their animals. Wolves are an ecologically important species, as well, as carcasses from wolf kills provide a food source for other scavenging animals.

Wolves are sexually dimorphic, with males being slightly larger than females, but wolves do not have a wide variance in their overall physical appearance. The domesticated dog, however, perhaps more than any other known species, domesticated or free-living, varies vastly in its size, morphological traits, and behavior, with the smallest dog on record weighing only 113 grams (.25 pounds) and the largest tipping the scales at a Rubenesque 294 pounds. As scavengers and predators, the domesticated dog has the blueprint of other predatory animals, including sharp front

---

[151] A "kickback" is a gesture where a chimpanzee lifts one hind leg behind them while they are quadrupedal. It is a universal gesture expressed across all chimpanzee communities.

canine teeth to catch and tear prey, long, lean, powerful muscles, front-facing eyes for binocular vision, a cardiovascular system capable of supporting both endurance and sprinting, fused wrist bones, and superior senses, specifically smell and sound, to detect their prey.

Externally, domesticated dogs look distinctly different from their wild wolf cousins, and they differ internally, as well. In addition to having a wide range of morphological varieties, from small to large, long-legged to short, slender to robust, dogs also have variation in their coat color and tail shape. Dogs have two kinds of coat varieties: single, with only a topcoat, and double coated, which have both soft down hair and a coarser guard hair. The latter coat originates from dogs who lived in colder climates and needed the added insulation to stay warm. Domestic dogs often exhibit the remnants of countershading, which is a common camouflage pattern found in free-living populations. Countershading is seen in dogs with white fur on their underside, chest, or face. The dog's tail can be thick or thin, straight, curled, cork-screw, or sickle-shaped. The primary function of the dog's tail is to communicate their emotional state. In comparison to wolves, domesticated dogs' tympanic bullae, which houses the sensory receptors, are small and compressed; dogs' brains are also 30% smaller than wolves', given expected brain size to body size ratios. Dogs' jaws, teeth, and eyes are also smaller than wolves', on average. Wolves do not have floppy ears, as many dog species do, and their paws are also twice the size of the paws of equivalently sized domesticated dog paws. Dogs also show great variation in their tails and coat colors, while wolves do not. The American Kennel Club recognizes eight distinct "groups" of dog breeds based upon physical, psychological, and behavioral characteristics. Some of these breeds are predisposed to certain types of diseases and ailments. Brachycephalic breeds often have breathing problems, German Shepherds are prone to hip dysplasia, Siberian Huskies are predisposed to autoimmune disorders, and Dobermans are predisposed to heart conditions. Dogs are also susceptible to parasites, including fleas, mites, ticks, heartworms, tapeworms, roundworms, and hookworms. Additionally, dogs are vulnerable to many of the health conditions that afflict humans, including neurological disorders, heart disease, diabetes, and arthritis. While the dogs' average lifespan varies greatly among the different breeds, the median lifespan is 10 to 13 years. A 2013 study found that dogs of mixed breed heritage lived, on average, 1.2 years longer than purebred dogs.[152]

Most domesticated dogs reach sexual maturity between six and twelve months of age. Female dogs have a gestation of approximately 63 days, with an average litter size, across the species, of five to six puppies, although smaller litter sizes are becoming increasingly common with purebred dogs. One of the downsides of selectively breeding dogs for so many generations is called "inbreeding depression." This occurs when closely related dogs are repeatedly bred to one another, and it decreases litter

---

[152] O'Neill, D. G., Church, D. B., McGreevy, P. D., Thomson, P. C., & Brodbelt, D. C. (2014). Prevalence of disorders recorded in dogs attending primary-care veterinary practices in England. *PLoS ONE*, https://doi.org/10.1371/journal.pone.0090501.

size and increases infant mortality rates. A study on dachshunds found that when the inbreeding coefficient increased, there were decreases in litter size and increases in the percentage of stillborn puppies.[153]

Dogs are highly intelligent animals with a particular ability to interact with humans, including learning the meaning of human words and sounds and being attuned to their gestural signals and even the odors associated with their health status. But this unparalleled ability to interact with humans does come at some cost; studies have shown that unlike their wild wolf cousins, dogs turn to humans to help them face their problems. A study on Australian dingoes (*Canis dingo*) indicated that dogs have lost a lot of their ancestral problem-solving skills, as dingoes outperformed dogs in non-social problem solving.[154] Studies comparing the problem-solving skills of domesticated dogs and wolves have revealed similar results, with wolves persevering to solve a task, while dogs looked to the human experimenters in apparent requests for "help."[155] Across testing conditions, wolves had an average success rate of 80% for solving a puzzle box game, while dogs averaged only 5% in both "alone" and "accompanied by a human" conditions. Other studies revealed that while wolves and dogs both have a social hierarchy, domesticated dogs will wait for the highest ranking dogs to go first, while the wolves all rushed in at the same time, despite the pecking order of their pack. In these studies, conducted at the Wolf Science Center in Austria, the dogs and wolves had to coordinate efforts to get a food reward. The food was in a tray that was attached to two ropes, and accessible only if the animals combined efforts. It appears as though dogs' cooperation has shifted from dog-dog interactions to dog-human ones. As the oldest domesticated species, though, it makes sense that dogs' behaviors would indicate an increased reliance on humans over their wild cousins.

Over the generations, dogs have learned to work together with their human partners, and the minds and behaviors of dogs and humans have been shaped by one another. This is evident in both what they struggle with (problem solving tasks, for instance) and where they excel (in reading and reacting to human communication). As aforementioned, dogs can outperform humans' ape cousins at certain tasks, such as referential pointing, and they demonstrate social-cognitive skills that parallel child development.

Humans communicate with their dogs, consciously and unconsciously, through a variety of sensory modalities, including verbal commands, gestures, touch, smell, and eye contact. Often, they are communicating information that they aren't even aware that they're transmitting, such as their health or emotional states, but dogs miss none of this. Not only can dogs detect our emotional states with their hearing

---

[153] Charlesworth, D., & Willis, J. H. (2009). The genetics of inbreeding depression. *Nature Reviews Genetics* 10(11): 783–796.
[154] Smith, B. P., & Litchfield, C. (2010). How well do dingoes, *Canis dingo*, perform on the detour task? *Animal Behaviour, 80*(1), 155–162.
[155] Udell, M. A. R. (2015). When dogs look back: Inhibition of independent problem-solving behaviour in domestic dogs (*Canis lupus familiaris*) compared with wolves (*Canis lupus*). *Biology Letters, 11*(9).

and vision, but they can also smell the difference between a person who is happy and a person who is stressed. Dogs communicate with one another with vocalizations, facial expressions and eye gaze, olfaction (pheromones, scents, and taste), and with their bodies, including posture of their body, head, and tail, and corresponding movements, including gait. And they communicate with their humans in similar ways, as well. A dog's auditory communication includes barks, growls, howls, whines, whimpers, screams, pants, and sighs. Vocalizations are context-specific and vary in pitch, amplitude, and timing that can potentially alter the meaning of the vocalization.

AS THE WORLD TURNSPITS

The dinner must be served at one;

Where's the vexatious Turnspit gone?

It's hard to imagine that prior to the advent of convection ovens and kitchens adorned with granite countertops and stainless steel accoutrements that dogs were a literal driving force in the cooking process. "Turnspit dogs" were bred for the specific purpose of helping their humans cook their meat to perfection, ensuring an even, steady rotation of a large roast or shank. These short-legged, long-bodied dogs would turn a wheel with a spit attached to it, rotating the meat over an open fire. Several hundred years ago, before we had modern kitchens, this was the preferred cooking method, and Turnspit dogs would run for hours to help their humans cook their dinners. Over time, other technology took the place of these hard-working hounds. By most accounts, the "Turnspit dog" is now extinct, while some believe that their descendants live on in a Welsh breed called the Bowsy Terrier. In the 1576 book *Of English Dogs,* this breed was referred to as the "Turnspete." Even a few centuries ago, people were aware that the Turnspit dog's days were limited.

The life of a Turnspit dog wasn't enviable; they were induced to run either by small food treats or by painful reinforcement. Hot coals would often be tossed into the dogs' wheel if their speed decreased below an acceptable level. While the suffering of the Turnspit dog ended long ago, their mistreatment later helped millions of other animals lead happier, healthier lives. In the 1850s, animal advocate Henry Bergh was appalled at the plight of the Turnspit dog. Their mistreatment was the catalyst behind the creation of the American Society for the Prevention of Cruelty to Animals (ASPCA).

NATIVE AMERICAN DOGS

Dogs first came to the new World approximately 10,000 to 14,000 years ago, crossing the Bering Land Bridge with their human companions. Humans and their dogs dispersed across the North American continent, remaining relatively isolated until explorers from Scandinavia reached Greenland in the 11th century. By the 1500s,

Europeans were widely exploring North and South America, and native dogs began to interbreed with the dogs that accompanied these explorers. Because horses aren't native to the North American continent, dogs often took their utilitarian place, helping First Nations peoples hunt, pull sleds, pack supplies, provide fur for clothing, and provide protection and companionship. In Northern British Columbia, the Salish Wool Dogs had plush coats that were sheared and used to create fabrics.

CHASING AMY

Dogs are classified under Carnivora, a diverse order that includes more than 280 species of placental mammals, including the suborder Caniforma, which includes canids, and the suborder Feliformia, which includes felines. Unlike cats, however, who are obligate carnivores and primarily (but not exclusively – a large percentage of cats can't seem to get enough catnip and many of them enjoy "grazing" on grass or sampling the occasional carbohydrate-rich indulgence) eat meat, dogs are omnivores who can thrive on a wide range of diets. Domesticated dogs often eat a combination of meat protein, grains, and vegetables. Having co-evolved with humans for so long, dogs' gastrointestinal systems differ from their wolf cousins: dogs have gene adaptations for starch digestion, while wolves do not. A study that analyzed the DNA of the remains of 13 ancient dogs tracked a gene involved with starch digestion. The remains, which originated from Asia and Europe and dated as far back as 15,000 years, revealed that between 4,000 and 7,000 years ago, dogs began to have more copies of the Amy2B gene. Wolves have two copies of this gene, while most dogs (excepting Huskies, which evolved among less agrarian societies and exhibit more wolf-like starch digestion capabilities) have between four and 30 copies of it. The Amy2B gene creates amylase, which breaks starch down into sugar. So why are we making our dogs eat grain? Early on during domestication, dogs were initially living with hunters and gatherers. As human civilization gradually moved toward more agricultural, sedentary lifestyles, meat became a smaller percentage of their diet. Farmers were now growing and eating higher percentages of wheat, millet, and corn, and in order to live alongside humans, dogs would need to become accustomed to eating these substances, too. Unlike every other domesticated dog breed, Huskies, like wolves and dingoes, only have two copies of Amy2B. Until very recently, evolutionarily speaking, Huskies have spent most of their time with hunters and fishermen, rather than those eating a more agriculturally-biased diet. Thus, their genetics – and their gastrointestinal fortitude – reflects a different evolutionary trajectory. The authors of the study stated: "This expansion reflects a local adaptation that allowed dogs to thrive on a starch-rich diet, especially within early farming societies, and suggests a bio-cultural co-evolution of dog genes and human culture."[156]

---

[156] Ollivier, M., Tresset, A., Bastian, F., Lagoutte, L., Axelsson, E., Arendt, M.-L., et al. (2016). Amy2B copy number variation reveals starch diet adaptations in ancient European dogs. *Royal Society Open Science*.

Dogs weren't the only species that experienced changes in their digestive systems with the rise of agriculture; the gut microbes of hunter-gatherers differ significantly from those of agriculturalists, too. A recent study compared the gut microbes of the Hadza of Tanzania, one of the last full-time hunter-gatherer populations, with those of residents from 16 industrialized countries. Microbes are inherited from one's mother, with no microbes present *in utero*; the first exposure to microbes is in the birth canal. Microbes co-evolve with us and vary depending upon one's diet and lifestyle. Thus, the digestive systems of dogs and humans were both changing to reflect their shared divergence from a hunter-gatherer lifestyle to a more sedentary agricultural one.

The rise of agriculture changed the digestive systems of domesticated cats, too. The ancestors of modern domesticated cats – as well as their contemporary wild cousins – eat diets that almost exclusively consist of meat. But during domestication, succeeding generations of cats were born with longer and longer intestines than their wild cousins. These longer intestines aided in the digestion of fibrous material; they were eating more grains, too. While they are obligate carnivores, those cats who chose to share their lives with humans needed to also be able to digest grain materials.

The natural history of a carnivorous or omnivorous species, like *Canis familiaris*, that has co-evolved with humans for tens of thousands of years, is going to differ from that of an herbivorous one, such as the horse (*Equus caballus*). Predators and prey experience the world with vastly different umwelts. What is salient to a predator species is not necessarily going to be salient to a prey species. While cats and dogs both share our homes, the dog has done so for three to four times as long – long enough, in evolutionary terms, to exhibit significant behavioral differences from both their feline housemates and from the last common ancestor that they shared with wolves.

And much has happened, evolutionarily speaking, since the ancestors of modern domesticated dogs diverged from the ancestors of modern wolves; so much so that Darwin, and countless other scientists and observers, have marveled at the morphological variation within this species. As man shaped dogs, though, they shaped him, in turn, resulting in increased rates of hunting success and protection from other predators, likely resulting in increased rates of reproductive success. The natural history of the dog, then, is just as much a story about canines as it is about modern humans.

From the onset of their partnership, the ancestors of modern man and modern dogs both benefited. Dogs had more reliable access to resources, including shelter and safety and food; thus, they were expending far fewer calories than their free-living counterparts. Those early dogs which exhibited the right combination of personality traits also had more opportunities to breed. When they hunted together, humans benefited from dogs' superior sense of smell, while dogs benefited from humans' sense of sight, including color vision and a higher vantage point. This was the start of a storied mutually beneficial relationship, with both species benefiting from the unique perspective of the others' umwelt.

All animals, including humans and dogs, view the world from their own unique umwelts. As primates, humans rely heavily on their sense of sight. But to cohabit successfully with humans, dogs learned, more than any other canine species, to maintain eye contact with people, as that was humans' preferred modality. Our canine companions, however, rely more upon their superior senses of smell and hearing. A dog's primary senses include olfaction, hearing, vision, taste, touch, and sensitivity to earth's magnetic field. A dog's world is chiefly navigated by olfaction, and approximately one-third of their brain is devoted to it. Recall the canine homunculus, comprised of a very large nose and oversized ears. In comparison to wild dogs, the domesticated dog has a high proportion of gene differentiation in the hypothalamus, which links emotional, endocrinological, and autonomic responses to exploratory behavior.

Dogs are not precocious (developing early and being independent at an early age, like many hooved prey animals are) when they are born; they're altricial, meaning that they are born in an underdeveloped state in comparison to most prey species. Unlike a prey species, like a horse, which can see, hear, and smell at birth and stands within an hour after being born, dogs come into the world deaf and blind and wholly dependent upon their mothers for many weeks. A dog's brain isn't fully developed until the age of two; prior to this time, they lack inhibitory control, much like a teenager. Dogs, like humans, have critical and sensitive periods of development. If they have missed out on socialization opportunities during these times of their development, they will struggle to understand their human family members.

Dogs are neurologically wired to detect the emotions of their human family members. During convergent evolution, dogs' brains have developed many analogous regions to human brains. To a certain extent, dogs experience many of the same emotional states that humans do, enabling them to relate to humans more. Dogs are able to detect changes in tone and emotional valence and they anticipate our reactions over perceived transgressions and achievements.

There are both costs and benefits to the human-canine relationship. Domesticated dogs benefit from the increased reliability of resources, including mate availability, shelter, and food; increased care of dependent offspring and decreased predation pressure. They incur costs from having to share their food with their human family members, as Ziva had to share with Alec, and from having less choice; they have to abide by the "rules" of another species.

Game theory, as it relates to biology and animal behavior, is crucial to understanding the development of the human-canine partnership. It has three main tenets. The first is that costs and benefits are conditional: the costs and benefits of an individual's social behavior are relative to the costs and benefits of another individual's social behavior. The second is that animals don't necessarily process numbers, at least not the way that humans do, so rules of thumb will arise from this. While the specific circumstances for an organism may be far different from the way they were when these behaviors arose, these rules of thumb will be deeply ingrained. The third is that there are iterative and non-iterative games. For example, an iterative game would be a dog visiting the same dog park every day at the same time and playing with familiar dogs each time. The dog would trust that other dogs would play fairly and return their toys as these would be dogs that they would be likely to interact with again. A non-iterative game would occur at a large city park where the dog had never visited before. There would not be a high likelihood of seeing these dogs again; thus, their interactions would vary widely between the iterative and the non-iterative scenarios.

The social structure of a species is based upon the distribution of resources within their environment. The social structure of the domesticated dog is unique from its closest relative, the wolf, as dogs have benefited from the reciprocal relationship with humans for millennia. For early dogs, cohabitating with humans meant easy access to food, shelter, mates, and potential care of dependent offspring. This impacted how they viewed the members of their household; while wolves have a hierarchy with the father wolf at the top, domesticated dogs view their families as a more amorphous grouping.

DOGUPATIONS

Historically, dogs helped humans hunt, pull carts, provide protection, herd, and even provide an emergency food source during difficult times. Today, dogs still provide many of these services, with the additional roles of being military and police dogs, search and rescue dogs, therapy dogs, and service dogs; some dogs have even traveled in space. Over the centuries, dogs have moved from living outside to living inside; most "pet" dogs live in the home now, bucking an historical trend that saw dogs living in their own doghouses through the 1960s and 70s. More often than not, those who share their lives with dogs refer to themselves not as dog owners, but as pet parents. These paradigm shifts have seen the commodification of the dog (pet parents spend $60 billion a year on their animals) and the related growth of pet-related businesses.

DOGS AND RELIGION

Religious views on dogs vary widely. In Chinese astrology, the dog is among the 12 honored animals and the second day of the Chinese New Year is celebrated as the "birthday of all dogs." It's customary to show kindness to all dogs on this day. In Catholicism, Saint Rocco is the patron saint of dogs. He was reportedly dying from plague when a dog came and licked his sores and brought him food, helping him recover from the disease. The Ancient Aztecs' god of death was a dog-headed monster named Xolotl, while the Ancient Egyptians' god of the underworld was the jackal-headed Anubis. A golden hound reportedly protected Zeus when he was a baby and the Hindus believe that the care and adoption of dogs can help them on their way to heaven. Under Jewish law, dogs and other animals are supposed to be fed before their human caretakers. Islam considers dogs to be "ritually unclean," however, so it's less common for dogs to be kept as pets in Muslim countries, although Muhammad did advocate kindness toward all animals, including dogs.

PUP CULTURE

Throughout time, dogs have held a special place in many cultures. In addition to being highly emblematic, with vivid depictions in paintings, carvings, and figurines over the course of human history, dogs have also been popular characters in literature and in film. In Homer's *The Odyssey*, Odysseus's dog, Argos, has been awaiting his return and recognized him after a 20-year absence. Dogs are protagonists in more contemporary books, as well, including *Lassie*, *Old Yeller*, and Nana from *Peter Pan*.

An overview of dogs on film over the past century is a microcosm of their natural history and reveals our changing viewpoints on them. In films like *Old Yeller* (1957) and *Lady and the Tramp* (1955), the dogs are beloved, but they're still placed outside of the home. As a puppy, Lady is first raised inside the house (but outside of the bedroom) and then moved to her own doghouse outside once she is an adult. Later films, such as *101 Dalmatians*, *Bolt*, and *Marley and Me*, depict dogs who share their lives inside the home. These subtle differences also reflect the dog's changing place in our lives; they were changing from guard dog and working dog to companion and family member. Whether they have a starring or a supporting role, dogs in film and literature are often depicted as heroic (*Where the Red Fern Grows*, *Turner and Hooch*, *Iron Will*, and *The Courage of Lassie*), sympathetic (*Shiloh*, *Eight Below*, and *The Incredible Journey*), and loyal (*Far From Home – the Adventures of Yellow Dog*, *Call of the Wild*, *Big Red*, and *Because of Winn-Dixie*). Very rarely are dogs depicted as villains, as was the dog in Stephen King's *Cujo*; it is the wolf who is saddled with the role of antihero throughout film and literature.

Dogs have inspired novels, films, art, and poems memorializing them. In 1981, movie star Jimmy Stewart shared his penchant for poetry – and just how much his dog, Beau, had profoundly impacted his life – on a poignant segment of the *Johnny Carson Show*. Stewart's poem, "I'll Never Forget a Dog Named Beau," is initially funny, noting his dog's disobedience and humorous characteristics. It begins:

He never came to me when I would call
Unless I had a tennis ball,
Or he felt like it,
But mostly he didn't come at all.

Stewart continues:

He knew where the tennis balls were upstairs,
And I'd give him one for a while.
He would push it under the bed with his nose
And I'd fish it out with a smile...

And there were nights when I'd feel him Climb upon our bed
And lie between us,
And I'd pat his head.
And there were nights when I'd feel this stare
And I'd wake up and he'd be sitting there
And I reach out my hand and stroke his hair.
And sometimes I'd feel him sigh and I think I know the reason why.

Then Stewart's poem strikes a somber note as it reveals that Beau has since passed on. Stewart wrote:

And there are nights when I think I feel that stare
And I reach out my hand to stroke his hair,
But he's not there.
Oh, how I wish that wasn't so,
I'll always love a dog named Beau.

"That stare" – a look that any dog lover can understand. A look that transcends time and the species barrier. You don't need to have met Beau to understand the profundity of Stewart's loss. Stewart was inspired to pen this poem shortly after Beau died; he received the phone call that his dog was near death when he was on a movie set in Arizona, and he left the set to be with his dog during Beau's final hours. Over the span of our partnership with them, our dogs have grown from outsiders to partners, and from partners to beloved family members. The segment memorializing Beau resonates with anyone, young or old, famous or unknown, who has ever loved and been loved by a dog. There is no other relationship quite like it. People often report that they grieve the loss of a pet – and in particular, their dogs – more than they do the loss of a human loved one. This should come as no surprise, given how our dogs love us unconditionally, from the moment that their eyes open until they last close.

Dogs today are living longer, aided by an awareness of the maladies that afflict them and knowledge about how to treat these issues. Their longevity is marked by the emergence of numerous fields, including animal palliative and hospice care, which is designed after the human hospice model. Rather than approaching end of life with fear and ignorance, organizations such as the International Association for Animal Hospice and Palliative Care (IAAHPC[157]) and the Animal Hospice, End of Life and Palliative Care Project (AHELP Project[158]), both founded in 2010, provide support for people whose animals are facing life-limiting illnesses or who have chronic conditions. These organizations teach pet owners what the end of life looks like, options for natural death and euthanasia, and physiological signs to watch for as pets approach the December of their days. There is increased awareness that dogs, like humans, require more care during their first and last years in life. In recent years,

---

[157] https://www.iaahpc.org.
[158] http://www.ahelpproject.org.

numerous other pet-related end-of-life businesses have been created, including aftercare companies that specialize in cremation and remembrances to memorialize pets who have passed away.

PULLING IT ALL TOGETHER

For practical purposes, modern animal behavior science suggests powerful concepts to aid in diagnosing and treating behavior issues, in all human-associated species, from captive exotics in zoos or kept as pets (such as parrots, ferrets, and various rodents) to livestock to companion animals such as dogs and cats. For instance, the modern science based on ethology suggests strongly that we need to assess and evaluate behavior in the animal's natural habitat, rather than viewing abnormal behavior in artificial or stressful situations, such as Schenkel's captive wolf studies from the 1930s and 40s. For dogs and other companion animals, the natural habitat is the home (and not the veterinarian's office, for instance) and we argue that applied animal behaviorists who do house calls (or park calls, or work calls, wherever the problem behavior exists) are far more successful than evaluations that are done within the clinic.

Modern animal behavior science clearly teaches us the importance of viewing the world from the animal's point-of-view; in some cases, literally dictating getting down to the dog's level to see what they see (or can't see) in, say, an overgrown dog park. We have to avoid anthropomorphizing our perceptions and emotions onto animals which do not have the same perceptions and emotions. This simple fact has converted so many clients to this approach, once explained; it makes good sense, and it works!

Modern animal behavior science shows us that there are species-typical behaviors and patterns of behaviors, for good, evolutionary reasons. Dogs are evolved to solve their challenges of survival with a high degree of social interaction and support. In addition, they are omnivores requiring both herbivorous and carnivorous diets and have evolved from ancestors who used a lot of movement in their hunting, cooperatively searching outs pockets of prey (say, rodents) or cooperatively, actively pursuing and capturing larger prey. Contrast this with cats, for instance: far less domesticated, far closer to their ancestral roots in behavior, and obligate carnivores (requiring a meat diet) as well as low-activity, wait-and-ambush predators, avoiding the use of social cooperation to a great degree. On the other hand, horses are large, all-day-foraging herbivores who, ancestrally, were prey species, evolving high degrees of alertness and social structure and cooperation to avoid predation, but not requiring complex behaviors in terms of finding food.

So what works for maintaining good mental health in dogs may not work as well for cats, or horses, and diagnosing and treating behavior problems is often quite different across species. And now, it appears that there are significant breed differences among dogs in behavior, as well: what works best for one breed, or at least a genetic breed group, might not work as well for another. A huge array of behavior, from learning modalities, sociability, use of dominance hierarchies, body language communication, exercise levels, and temperament may be quite different between, say, wolf-like "ancient breeds" and so-called "modern European breeds." The well-qualified owner, trainer, handler, or veterinarian, and certainly most qualified behavior specialists are aware of these factors, perhaps at an even subconscious level. But it is well-supported by modern animal behavior science.

One of the most dramatic practical examples of the co-evolutionary connections between humans and dogs is the use of assistance dogs in the courtroom. JCH has had the honor of serving on the advisory board of the Courthouse Dogs Foundation since its inception, which has played an active role in the movement to bring assistance dogs into the legal realm. Its greatest application has been in the use of these dogs, usually a Labrador or Labrador-Golden Retriever hybrid, in the forensic interview process for minor-age victims of sexual abuse. The trauma that these children have experienced is horrific, but the need for them to repeat their stories and thus, relive that trauma, is necessary in our justice system. It's becoming clear that the presence of a dog during these interviews can significantly reduce the mental health trauma of the experience, both in the short-term as well as the long-term (Fig. 43).

Should we be surprised by this? We don't think so; as we have discussed throughout this book, the lengthy co-evolutionary history between human and dog has resulted in significant abilities to read body language, to act empathically (in both directions), to sense emotions, and to respond with appropriate comfort behavior. The gifted scientists who manage the best assistance dog programs have used the sciences of genetics and behavior to develop information and assessments which have amplified these traits (already strong in Labradors and Golden Retrievers from their development as hunting companions).

Therapy dogs have provided support and solace in a variety of settings. College Dogs, which was founded in 2010, works with students at universities and colleges in Washington State and with patients at nursing homes. Founder Laurie Hardman has seen the profound impact that these therapy dog teams can have, particularly after a stressful incident.

"Following a crisis at one of the schools – one of them was a suicide, and there were two different shootings – we saw that there was a 'desperate calm,'" she recalled. "Students and staff alike were looking for that calm, sweet, 'ask nothing of you,' unconditional love from the dogs. These interactions are quieter on the part of the people than our typical therapy dog visits, and people use lower tones and have slower conversations. There's calmer petting. The dogs 'pointedly connect' with them; whereas the dogs may not do 'nose to nose' interaction normally, they're truly trying to make an 'I'm with you' connection with these people. They do this after something traumatic has happened to the people they're visiting. It's a seemingly intuitive knowledge that 'something is different;' a need to check in and ask, 'Are you okay?' The other reaction that I see is modeling the calm behavior of the students and staff. The dogs totally relax against and with the students. They gently nudge them for petting sometimes, but it's a more physical, calm connection.

"The dogs 'bare' themselves completely, and the students can obviously be seen physically relaxing, slowing down, and making connections with the dogs. The people

FIG. 43

Courthouse Dogs at work.

connect with the dogs, and the dogs connect with each person, acknowledging each new person. I see the dogs offering low-level calming signals, such as an occasional yawn, and more often than not, the students yawn back, without even realizing what's taking place. The dogs will offer their tummies and backs. The students will pet them very calmly, slowly, and methodically; more so than during a 'normal' session. They also interact with one another as they sit around each dog and pet it, moving about the room to pet each dog. The students also appear to be more aware of each other during these visits."

For more than 15 years, Hardman has visited weekly at Northwest Hospital's Gero-Psych ward, alternating between her two therapy dogs, senior Billie and puppy

Fido. "During these visits, I spend 30 minutes in two locked units, though often, I go to individuals in the 'quiet' room. A lot of the focus in these visits is to stir memories and to orient," she explained. "A physical interaction can be almost like an electrical zap to these patients. Somebody who seems completely 'shut down' might just watch at first, but then at a later visit, reach out and touch. And then the light bulb flickers and conversation starts."

During one of these visits, Hardman took her puppy, Fido, who was then only nine weeks old. "I walked past the 'quiet room,'" she recalled, "and a woman came running out and said, 'Oh, please, can you put your puppy on my father's bed?' Staff nodded 'of course,' so I looked him over, saw that he had no physical issues to be concerned about, and placed Fido on him. The moment I placed Fido in the crook of his arm on the bed, he turned his head down, smiled, and totally focused on Fido, and then…he began talking to him." This man had been nonverbal, and only making non-sensical, loud outbursts since he was admitted.

"When I came back the next week with Fido, this gentleman was no longer in the quiet room. Once he saw Fido, he said, 'Hi, Fido!' and his daughter started crying that her father remembered Fido's name. We put a pillow on her father's lap, placing Fido on the pillow. Fido looked up and started licking his chin. The gentleman laughed and said, 'I had a cookie!' The connection was unbelievable."

With their innate ability to read human gestures and react to their emotional states, it's easy to see why dogs are the go-to animal for therapy needs. What's harder to understand is how we're continuing to manipulate this species, to the detriment of the dogs themselves. Over the course of only 100 years, breeds created by human selection have been profoundly altered from their prior appearances. In 1915, the German Shepherd was considered to be a medium-sized (approximately 40 pounds) dog with a back that was parallel to the horizon. Today, most German Shepherds weigh at least 30 pounds more, and they have backs that slope (dramatically), with a decline from the shoulder to the tail. As a result, they have issues with hip dysplasia. The bull terrier is unrecognizable from the dog with the same name from 100 years ago; it now has a very convex nose in comparison to the smaller-faced founding members of their breed. The Bassett Hound, too, has been altered; one century ago, its legs were longer, its ears shorter, and it had fewer skin folds; this breed now suffers from dermatitis from this excess skin and eye issues with its "droopy eyes." The last century hasn't been kind to brachycephalic breeds in particular. The pug has experienced serious genetic modification over the past 100 years; most now suffer from high blood pressure, decreased oxygen levels, breathing problems, dental issues with their truncated faces, and dermatitis in their skin folds. Their signature curly-cue tail is problematic, too; it's actually a genetic defect that can potentially cause paralysis later in life. Boxers, another brachycephalic breed, are prone to issues with thermoregulation and breathing. Even Saint Bernards are now more brachycephalic (and much larger, too), than they were 100 years ago. They are susceptible to eye and eyelid issues, Stockard's paralysis, and bleeding and clotting disorders. Salukis are now so lithe that they don't resemble the Salukis of 100 years ago, and suffer from heart defects. Dachshunds are less proportionate than their great-great grandparents were; they often suffer from intervertebral disc disease and progressive retinal atrophy. Humans are largely driving the direction of dog evolution, and as stewards of this species, it's our responsibility to limit harmful traits, spay and neuter them to decrease population growth and corresponding shelter euthanasia rates, and provide the best possible care for them that we can.

Dogs have a unique place in the animal kingdom. They have shared their lives with humans for tens of thousands of years, but they are unlike any other domesticated species. Dogs are social carnivores, not grazing prey species like horses or cows or small, asocial carnivores like cats. Descended from the ancestors of wolves and shaped through artificial selection by man, dogs differ from all other species in their ability to intuit and respond to their human companions. The natural history of the dog is just as much a natural history of modern man. Were it not for their support, guidance, and companionship, we would not be the species that we are today, and neither would they. Through choosing to partner with each other, we have also changed the direction of one another's evolution. While dogs have helped us carry our belongings, track down and capture prey, keep us warm in the cold, defend us from aggressors, locate the injured and lost, and perform day-to-day activities like locomotion, they are now moving into more subtle and closer actions, like detecting the onset of seizures, of illnesses like cancer, and even to respond appropriately to significant emotional disturbance. And yet, domestic dogs remain outside the realm of modern science exploration: We know more about communication in birds and lizards than we know about the communicative abilities of dogs, we know more about sensitive periods and windows of influence in ducks than we do in dogs, we know more about the genetics of behavior in fruit flies than we do in dogs, and we know more about space-guarding (territoriality) in monkeys than we do in dogs.

Only one century ago, most domesticated dogs were working animals, but today, they are family members, leading longer, healthier lives, due in large part to advances in veterinary medicine and a greater awareness of their species-specific needs. We're now able to screen for and treat aging-related issues, including cancer, we now have pet insurance, and we even have devices to help our disabled pets lead fuller lives. In a relatively short amount of time, dogs have gone from the outside fringes of our caves to the inner circle of our families. Their presence, appearance, and behavior are a reflection of our own shifting values and beliefs.

Darwin was one of the first scientists to see the value in studying domesticated dogs, but until very recently, their study has gone out of vogue. In JCH's opinion, dogs have been considered an "un-natural" species, not an appropriate subject for the study of evolutionary biology, and to some extent, until we *really* understood the (heavily human-manipulated) genetics of dogs, this was true.

Our hope is that, with the advent of modern DNA methods, and our recent unveiling of the *real* derivation of dog breeds (not the entertaining, but frequently incorrect breed origin stories), we will begin a real effort to apply the modern science of behavior, and genetics, physiology, and medicine, to learn even more about our partner in evolution, the domestic dog.

Conclusion

Twilight was falling and a net of stars cast out above Alec and Ziva's blind. They sat silently, Alec staring down at the watering hole and Ziva's nose to the air. Ziva's head began to jerk upward quickly as she chased something that Alec couldn't see or smell. Then he saw it – motionless, reddish brown against the green foliage. He tapped Ziva's shoulder, unclipped her leash, and pointed; she followed the direction of his finger and then locked her eyes on the clearing, just to their ten o'clock. There, at the edge of the brush, was a young buck moving tentatively. He paused as he entered the clearing and then walked to the water's edge, taking a long drink. Alec watched as his throat flexed with each gulp of water that went down his neck. He slowly raised his weapon. Ziva stood at attention, her mouth open, tail up, ears up, watching Alec, and then the buck, and then Alec again.

The first shot sailed out from the blind; for one moment, they waited, unsure if it would land the buck. And then the shot struck the buck's head – a lucky shot, really, Alec thought in disbelief. Dazed, the buck stumbled forward and fell into the water face first. He struggled in the shallows, floundering, unable to regain his feet. Alec sprang from the blind, but Ziva was quicker; her powerful haunches propelled her down the slope. She had the buck by the time Alec reached the water's edge. The buck struggled in her grip, but one more shot and he was gone.

As Alec cleaned the deer, he watched Ziva with a smile. She was licking her lips, drooling, looking from the carcass and then back to Alec. She sniffed and then sneezed as Alec continued to cut the skin into long strips. She yawned as he pretended to not notice her keen interest in the buck, and then she leaned back, raising a front paw. She watched every movement of his hands as he took the liver from the body. Alec laughed quietly and threw it to her. She eagerly lay down, ripping out chunks of warm meat while Alec dressed the buck for the hike back to their truck. Once he had quartered the animal, he placed the pieces into his duffel and into a bag that he unfolded and attached to Ziva. The animal was too large for him to carry out on his own and Ziva's pack was particularly useful during the unusual occasion when they had such a big kill.

The rising half-moon glittered on the path as they walked back to their vehicle at the trailhead. Alec loaded the body in the back and then they took the truck the rest of the way back to town. Ziva buried her nose into his stomach again, this time nipping pointedly at his shirt.

"Ouch!" Alec said, brushing her head away as she pinched a fold of his stomach with her sharp teeth. "What are you doing? Stop!" Ziva stared at him fixedly.

As Alec and Ziva got closer to their neighborhood, he noticed all of the lights on in the view homes overlooking the sea. So many new homes – so many new neighbors. Dogs began to bark in the distance, and Ziva stared toward the sound, uttering a low *urrrr*, but not barking aloud. She glanced over at Alec, who patted her thoughtfully. It seemed like all of the new neighbors had dogs, and lots of them.

Alec rounded the truck down the final descent into their drive, smiling at all of the lights that had been left on for them. He pulled the truck into the drive and then unloaded the meat and hide. Before he could even get in the doorway, two children, one small and one almost grown, greeted Alec eagerly, and then embraced Ziva, who whined and bounced and rolled onto her back, ready for them to give her belly rubs. He laughed, brushing off his boots and taking off his deerskin jacket, shaking off a light dusting of snow. He watched it glitter down into a pool of water in his front yard. He soberly stared at the reflection that met him – prominent brow, bulbous nose, small chin, large forehead, and long, auburn hair already streaked with white. He looked so different from the slight, light-eyed man that he had seen in the forest. He looked so different from all of the new people coming to his quiet seaside town.

He patted Ziva on the head as his wife, Telin, and his sister-in-law, Hulya, greeted him at the mouth of the cave. The torches lined along the walls bounced off of the limestone, creating a greenish gold glow that illuminated their faces. He asked where his brother was, but they shook their heads – Alec was the only one who had returned from hunting so far. Ziva bounded into the back of the cave, eager to greet the two other dogs in her family, and Telin and Hulya excitedly opened the deer hide sacks containing the meat that would help sustain them for another month. Alec's young son ran out into the clearing before the cave to retrieve the hand truck, pulling the oversized sled over toward the blazing fire. Alec's daughter, who was almost grown, stared out of the cave wonderingly as she heard the sounds of distant, unfamiliar voices.

Alec walked slowly to the back of the cave, tracing his hand along the grooves that his ancestors had carved into the rock for generations: grid patterns, cut masterfully into the dolomite. He rubbed his thumb and forefinger along his weathered slingshot and glanced back out as he heard his neighbors' dogs bark once again. Snow was falling heavily at their doorway, but here, within the cavern that would carry the moniker "Gorham's Cave" 25,000 years later, Alec, his family, and his dogs spent a quiet evening together – much like countless other families would, for millennia afterward.

Afterward

Alec and his family would be the last Neanderthals to occupy Gorham's Cave, a natural opening in the Rock of Gibraltar's southeastern face that was "discovered" by modern humans in 1907. During that year, Captain A. Gorham of the Royal Munster Fusiliers spied the fissure and inscribed his own name and date on the cave's wall, tens of thousands of years after *Homo neanderthalensis* disappeared from the planet, leaving after *Homo sapiens* as the only extant *Homo* species. Did Captain Gorham see the ancient, abstract carvings when he claimed the cave as his own? While Gorham's Cave didn't have fossil footprints like Chauvet Cave, did he wonder who had passed before him as he wondered at the 39,000-year-old hash mark scratches on the cave floor, and wonder who to attribute them to? Did he find tools, fossils, or evidence that Neanderthals hunted game from the land, air, and sea?

For more than 100,000 years, Neanderthals occupied the caves of the Gorham Complex. Between 25,000 and 30,000 years ago, Neanderthal populations were effectively driven out of their last strongholds, and most scientists believed that they disappeared from the gene pool. Recent genetic studies reveal that Neanderthals weren't truly "eradicated," though; they were "absorbed" into the lineage of modern man. When they coincided, Neanderthals and humans interbred for several thousand years. Curious individuals, just like Alec's daughter, likely had dwindling partner choices outside of their own families, and ventured out into the settlements of the "other" people to find mates. Most Europeans and Asians have between 1 and 4 percent Neanderthal DNA, while indigenous sub-Saharan African people have none. Neanderthals migrated through Europe and East Asia, but not Africa, and Neanderthals interbred with *Homo sapiens* on those continents, revealing that the shared existence between the two groups perhaps wasn't *completely* antagonistic.

When Alec's family lived in the cave, it was a time of great change: the climate was shifting, much as it is today; populations were in flux, much as they are today; and the human-canine relationship was redefining how man – *Homo neanderthalensis* and *Homo sapiens* alike – was existing. Twenty-five thousand years ago, the cave would have still been at least a mile from sea; for the cave's first inhabitants, fifty-five thousand years before common era, the earth was still in the grip of the last ice age, and the cave was a good three miles from the Mediterranean Sea. Fifty-five thousand years ago, game was still abundant; including the megafauna of the last ice age; so too were Neanderthals abundant, while *Homo sapiens* was far less common, up until the point that they shared time and space with people like Alec's family.

Today, Gorham's Cave Complex is only meters from the Mediterranean, and it's the only UNESCO World Heritage Site in Gibraltar. Gorham's Cave, along with Bennett's Cave, Hyaena Cave, and Vanguard Cave, comprises the Gorham's Cave complex. While there are numerous hypotheses about why Neanderthals eventually died out, a popular hypothesis postulates that humans eradicated their Neanderthal rivals with the help of early dogs that were bred from the ancestors of wolves. Whatever the selective advantage was, it's clear from both behavioral and fossil evidence that the partnership between humans and dogs is an ancient one. Who's to say that Neanderthals didn't also tame the ancestors of wolves and cohabit with these early dogs? Perhaps *Homo sapiens* just tamed more dogs, or their dogs had higher survival rates, thus helping them hunt more game and win more battles than his Neanderthal cousins; perhaps he also had more adaptive technologies, a heartier immune system, and adapted better to climate change. Just as sickle cell anemia can protect some against malaria, perhaps the population dwindled and then individuals and families who once had selective advantages now had recessive – and dangerous – mutations. Neanderthals' more robust bodies, in comparison to *Homo sapiens*, were considered to be particularly adapted to the cold; perhaps as the last ice age released its grip, they became less fit for the environment. Recent findings with a 49,000-year-old Neanderthal from El Sidron, Spain – 626 miles from Gorham's Cave – also revealed a potential genetic cause for the decline of the Neanderthals: only female Neanderthals and male *Homo sapiens* could have produced viable offspring. Female *Homo sapiens* were genetically incompatible with male Neanderthals due to a divergence in their Y chromosomes. This could have led to the miscarriage of their male fetuses and decreased reproductive success. According to Stanford University's Carlos Bustamante, the Neanderthal Y chromosome has never been detected in any human sample ever tested.

Whatever the reason, dogs have shaped the path of human evolution, just as much as humans have shaped the path of canines. Humans have created almost 400 distinct breeds of dogs within *Canis familiaris* and dogs, in their great variety, have supported the success of *Homo sapiens*. Dogs can help us hunt, provide therapy and support, protect us, guide us, and detect abnormal cell growth, illegal substances, and explosive devices. Ziva could already smell the cancer growing in Alec's stomach that would claim his life before the passage of another year. While cancer in prehistoric man was likely more uncommon than it is in the modern era, there were also no known treatments for internal maladies that couldn't be diagnosed. Twenty-five thousand years later, dogs can detect abnormal cells, and modern technology can treat these diseases.

It's unknown what perfect storm of factors gathered together to eradicate the majority of the Neanderthals – it was probably a synchrony of environmental, physical, genetic, and social factors, similar to the extinctions of canine species such as the thylacine. And it's also not known just how much *Homo neanderthalensis* coexisted with canines, but from what we know of modern man and modern dogs, it wouldn't be surprising if they, too, shared their hearths and homes with man's first friend.

Index

Note: Page numbers followed by *f* indicate figures.

A

Adaptation, 6
Adenine, thymine, cytosine, and guanine (ATCG), 7
Aggressive behavior, 77, 157–159, 162–163
Akitas, 162–163
Allogrooming, 22
Alloparental care, 124
Altruism, 141–142
American Cocker Spaniels, 162–163
American Society for the Prevention of Cruelty to Animals (ASPCA), 164, 181
American Veterinary Medical Association (AVMA), 134–135, 164
Anatomy and Philosophy of Expression (1824), 96–97
Anderson, David, 97
Anecdotal cognitivism, 37–39
Animal Cognition, 153
Animal emotion denialists, 93
Animal Hospice, End of Life and Palliative Care (AHELP Project), 187–188
Anthropomorphism, 33–35
Antidepression medications, 101–102
Anxiety, 75, 117, 151, 168–169
Apes, 42
Applied Animal Behaviour, 153–154
Auditory communication, 58–59
Audubon, John James, 172
Australian Cattle Dogs, 162–163
Australian dingoes *(Canis dingo)*, 180
Australian shepherds, 32, 70–71
Automatons, 93
Aversive-based training, 77
Avians. *See* Birds
Axelrod, Robert, 142–143
Axon, 64–65

B

Barking, 59, 181
Base emotions, 93–94
Bat, 41, 41*f*
Beach dogs, 75
Beagles, 162–163
Bee, 41, 41*f*
Beetles, 82
Begging behaviors, 49–51, 55
Behavioral ecology, 170
Behaviorism, 72
Behaviorist learning theory, 72
Bell, Charles, 96–97
Belyaev, Dmitri, 126–127
Belyaev fox, 161
Bernese Mountain Dogs, 162–163
Birds
　adaptations, 6
　descent with modification, 6
　finches and mockingbirds, 1–3, 4*f*
　imprinting, 37–39
　pigeons, 3–4
　selective pressures, effect of, 6
Biston betularia. See Peppered moth
Black Plague, 122, 123*f*
Bloodhounds, 5, 11, 87
Body language, 49, 52–53, 55, 130, 189
Bonobos, 33, 124–125, 155–156
Bottlenose dolphin, 124–125
Bowsy Terrier, 181
Boxers, 191–192
Brachycephalic breeds, 179–180
Brain, 63–64
　cat, 65–66
　central nervous system, 65
　cerebellum, 67–68
　cerebral hemispheres, 66–67
　dog, 67*f*, 85
　　average weighs, 65–66
　　developmental phases, 68–71
　　encephalization quotient, 65–66
　　frontal lobe, 66–67
　　homunculus, 67
　　and humans, communication between, 75–76
　　learning, 71–73
　　olfactory portion of, 66–67
　　sensitive and critical periods of development, 73–75
　　training, 77–78
　　and wolves, neurological and behavioral distinctions, 68
　elephant, 65–66
　forebrain, 65
　functional parts of, 65
　gray matter, 64–65
　gray wolf, 65–66
　hindbrain, 65
　human, 63–67, 64*f*, 66*f*, 85
　midbrain, 65
　neurons, 64–66

Brain *(Continued)*
  occipital lobe, 67–68
  parietal lobe, 67–68
  reptile brain, 66
  white matter, 64–65
Breed-specific legislation (BSL), 163–164
Brittany, 162–163
Brown capuchin monkey *(Cebus apella)*, 147
Bulldog, 5

C

Calming signals, 59
Caniforms, 8–9
Canine Behavioral Assessment and Research Questionnaire (CBARQ), 162–163
Canines. *See* Dogs
Canis familiaris. *See* Dogs
Canis lupus. *See* Wolf
Canis species, 8–9
Carta marina, 63–64
Cartesian dualism, 37–39
Cats, 18*f*, 20*f*
  agriculture, 183
  behavioral issues, 21–22
  brain, 65–66
  depression in, 101
  vs. dogs, 20–21
  genetic differentiation, 19–20
  hearing, sense of, 82
  humans, relationship with, 17–19, 22
  interspecific relationships, 34–35
  obligate carnivores, 20–21
  obligate carnivores, 182–183, 188
  selection pressures, 19–20
  smell, sense of, 82, 85
  taste, sense of, 81, 91
  vibrissae, 88–89, 90*f*
  witches, companions of, 19
Cattle, 136–137
Causation, 39
Centers for Disease Control (CDC), 164
Central nervous system (CNS), 65
Charles II, 26, 27*f*
Chauvet footprints, xviii*f*, xvii, xvii–xviii
Chauvet-Pont-d'Arc Cave, 175
Cheese, 57
Cheetahs, 25–26
Chihuahuas, 7–8, 113–114, 162–163
Chimpanzees *(Pan troglodytes)*, 49–50, 94, 95*f*, 155–156, 159–160
  and bonobos, 33
  encephalization quotient, 124–125, 128–129
  humans, xix–xx
  visual gestures, 55
Chow chow, 12–13
Cilantro, 80–81
Classical conditioning. *See* Pavlovian conditioning
Cockapoo, 11
Cognitive emotions, 98–99
Cognitivism, 72
College Dogs, 189
Color vision, 80
  dichromatic vision, 83
  humans, 80
  red-green differences, 83–84
  reflected colors, 80
  tetrachromatic vision, 83
  trichromatic color vision, 83
Communication, dogs, xix, 46–47, 75–76
  auditory communication, 58–59
  body language, 49, 52–53, 55
  calming signals, 59
  eyes and ears, 50–51
  growl, 53, 59
  head, position of, 52
  licking, 51–52, 56
  play behaviors, 53–54
  play-bow position, 52–54, 52*f*
  referential communication, 47
  submissive position, 51–52
  tail movement, 49–50
  yawning, 51–52
Conditioned response, 72
Constructivism, 72
Contagious diseases, 122
Contest competition, 156
Context-specific referential communication, 47
Convergent evolution, 17, 127
Cooperative breeding hypothesis, 123
Coprophagy, 57
Copycat/tit-for-tat strategy, 144–145
Costs and benefits of being social
  "alloparenting," 123
  behavioral traits, 132–133
  canine-human relationships, 134
  canine parvovirus, 122
  cooperative breeding hypothesis, 123
  domestication, 126–127, 136–137
  heritability, 133
  on individual level, 124
  interspecific social relationship, 121
  judgment bias, 134
  kin selection, 129–130
  personality, 132–133
  predation pressure, 124

reciprocal altruism, 130
resources, 122
social behavior and genetics, 121
social group, 129–130
social living, 122
social structures, 124
solitary animals, 121–122
on species level, 124
temperament, 132–134
Countershading, 179–180
Courthouse Dogs Foundation, 189, 190*f*
Coyotes *(Canis latrans)*, 7–9, 54, 173–175

D

Dachshunds, 162–163, 191–192
Darwin, Charles, 1*f*, 37–39, 141, 175, 192
  Darwin's Finches, 1–2
  dog aggression, 157–158, 158*f*
  dogs
    affinitive behaviors, 96–97
    emotion, 96–97, 97*f*
  nonhuman animals, emotional lives of, 93
  theory of evolution (*see* Evolutionary theory)
  The Voyage of the Beagle, 2–3
Dawson, Charles, 171
Debunking dominance theory. *See* Dominance theory
de Malebranche, Nicolas, 96–97
Deoxyribonucleic acid (DNA), 7
Descartes, Rene, 93, 96–97
The Descent of Man (1871), 96–97
Devil-Jack Diamond fish, 172, 172*f*
Dholes, 7–8
Diamond, Jared, 113–114, 124–125
Dichromatic vision, 83
Dingoes, 7–8
Doberman pinschers, 179–180
Dog Facial Action Coding System (DogFACS), 44–46
Dogs, 4, 5*f*, 85, 156–157
  adaptations, 6
  artificially selected traits, 23
  bloodhound, 5, 11
  brachycephalic and nonbrachycephalic dogs, 30–31, 31*f*
  brain (*see* Brain)
  bulldog, 5
  vs. cats, 20–21
  cheese eating behavior, 57
  communication, 181
  convergent evolution, 17, 184
  coprophagy, 57
  depression, 100–102
  descent with modification, 5, 16–17
  designed dogs, 11
  divergent socioecology of, 156–157
  encephalization quotient, 128–129
  exaggerated characteristics, 11, 30–31
  facial expressions, 44–46
  formal dominance, 167
  garbage dump hypothesis, 13, 15
  grass eating behavior, 57–58
  gray wolves, ancestors of, xix–xx, 7–8, 14–15
  hearing, sense of, 43–44, 87–88
  herding dogs, 23, 32
  howling, 56
  humans, relationship with, xx, 13, 17
    anthropomorphism, 33–34
    Chauvet footprints, xviii*f*, xvii, xvii–xviii
    family members, xviii–xix, xxi
    flehmen response, 57
    food reward, 36
    food source, 12–13
    guilty/dog shaming, 33, 35–36
    high-risk behaviors, 36
    hunting partners, 16–17
    lupomorphizing, 33
    one/two-legged bicycle response, 57
    pet parents, xviii–xix, xxii–xxiii
    protectors, 16–17
    verbal and nonverbal communication, xix, 46–55, 58–61
  inbreeding, 23, 25, 29–30
  interspecific relationships, 34–35
  interspecific reproduction, 8–9
  intraspecies reproduction, 8–9
  Italian greyhound, 5, 11
  maternal DNA, 14–15
  mixed-breed dogs, hypoallergenic, 11–12
  mixed-breed heritage, 179–180
  morphological changes, 11
  natural history
    Amy2B gene, 182–185
    big fat rotten cows, dragons, and piltdown man, 171–173
    binomial nomenclature, 176–177
    dogs and religion, 185–186
    dogupations, 185
    evolutionary process, 176–177
    feral children, 176
    historia naturalis, 173–175
    modern animal behavior science, 188
    Native American dogs, 181–182
    pup culture, 186–188
    social structure, 185

Dogs *(Continued)*
   therapy dogs, 189
   wolves *vs.* dogs, 177–181
  olfactory cues, 41, 55
  opportunistic omnivores, 20–21
  pet hypothesis, 13, 15
  phenotypic variation, 7–8
  pug, 5
  purebred dogs, heritable conditions, 23–25
  rolling behavior, 57
  sensory experience, 43, 86
  shelter dogs, 22
  smell, sense of, 42–44, 59–60, 86, 91–92
   accuracy, 81
   diseases, detection of, 86–87
   drugs and explosives, 42–43
   human-related scents, 85
   long snouts, 87
   olfactory memory, 85–86
   olfactory receptors, 43, 85
   pain and fear, detection of, 86–87
   sickness, detection of, 86–87
   three-dimensional, 42
   vomeronasal organ, 43
  social carnivores, 192
  specialized dog breeds, 29, 30*f*
  tail chasing, 56
  taste, sense of, 81, 91–92
  umwelt, 41, 84
  vibrissae, 88–89, 90*f*
  vision
   blind corridor, 84
   color blind, 83
   dichromatic vision, 83
   dog park designing, 84
   low-resolution vision, 43
   red-green differences, 83–84
   superior vision, 43
  wolf dogs, behavioral issues, 9–11
  yawning, 105–106, 106*f*
Domesticated dogs. *See* Dogs
Dominance theory
  aggression, 158*f*
   aggressive actions, 158
   environmental and genetic factors, 161
   ethology-based types of, 157–158
   fear-anxiety and redirected aggression, 158–159
   husky mix, 160
   leash aggression, 160–161
   physiology of anger and, 158
   resource guarding/protective aggression, 158–159
   social hierarchy, 159
  animal behavior, 166
  anxiety- and fear-based behaviors, 169
  aversive training, 168–169
  breed-specific legislation, 163–164
  confrontational techniques, 167
  despotic social system, 166
  mice and humans, 162–163
  outdated "pack theory" ideas, 168
  personality trait, 166
  positive reinforcement methods, 170
  RHP, 170
  social behavior, socioecology of, 155–157
  social dominance/alpha model, 165
  social system, 166
  "The Dog Whisperer," 164–165
  "the pecking order," 166
Dopamine, 97

E

Emotions
  age of denialism, 93–94
  aggressive behavior, 107
  anthropomorphism, 108
  arrested development, 103–105
  behavioral indications, 107
  behavior modification methods, 108
  body language, 107
  canine depression, 100–102
  coevolutionary relationships, 108
  cognitive function, 107
  communication, 105–107
  contagious yawns, 105–106, 106*f*
  definition, 95–96
  depression
   clinical signs of, 103
   social interaction, 103
  facial expressions, 98
  fear and anxiety signals, 108
  functional magnetic resonance images, 105–106
  genetic similarity, 98
  internal endocrine state, 105–106
  MNS, 105–106
  paradigm shifts
   animal emotional expressions, 96–97
   dopamine receptor molecules, 97
   expressions, 96–97
   flies and humans, 97
   human emotional expressions, 96–97
   hyperactivity, 97
   theory of mind, 94
   tool manufacture and modification, 94–95

pheromones, 107
physiology and biology of, 98–100
visual and chemical cues, 106–107
Encephalization quotient (EQ), 65–66, 124–126, 125*f*
Endemism, 2–3
Epigenetics, 23
Ethograms, 47–50, 48*f*, 52–53
Ethology, 36–40
Ethylene glycol, 81
Eusocial insects (Hymenoptera), 113, 114*f*
Evolutionarily stable strategies (ESS), 150
Evolutionary game theory (EGT), 149
Evolutionary theory
    birds
        adaptations, 6
        descent with modification, 6
        finches and mockingbirds, 1–3, 4*f*
        pigeons, 3–4
        selective pressures, effect of, 6
    dogs (*see* Dogs)
    moths, industrialized melanism with, 6, 7*f*
Expression of Emotion in Man and Animals, 93–94, 96–97

F

Facial Action Coding System (FACS), 46
Facial expressions, 44–46
"Fear grimace," 159–160
Feliforms, 8–9
Felines. *See* Cats
Fennec fox, 7–8
Feral dogs, 73
Finches, 1–3, 4*f*
Flavor blasting, 91–92
Flehmen response, 57
Food reward, 36, 39, 72
Foxes, xx, 10–11, 10*f*
Functional magnetic resonance imaging (fMRI), 88

G

Gage, Phineas, 63–64, 64*f*
Galápagos Islands, 1–3
Galápagos mockingbird, 1–3, 4*f*
Galaxy, Jackson, 21
Game theory
    dogs
        costs and benefits, 150
        equitable outcomes, 148
        stress and aggression, 151
        transient dogs, 151
        unexplained/unexpected behaviors, 150

Game theory
    biology and animal behavior, 184–185
    canine brainpower, 153–154
    conditional interactions, 140, 151
    dear-enemy effect, 145–146
    defense-of-territory behaviors, 145–146
    domestic dogs, 147
    "How I behave depends upon how you are behaving… (maybe!)"
        animal behavior, 148
        behavior issues, 152
        EGT, 149
        ESS, 150
        Hawk-Dove game, 150
        ravens, 152–153
        reciprocal altruism, 151
        shareable resources, 150
        social behavior issues, 148
    "inequitable payoff," 147
    miscommunication, 146
    monkeys, 147
    Nash equilibrium, 140
    parlor games, 139
    "prisoner's dilemma"
        altruism, 141–142
        animal cooperation, 142–143
        cooperation, evolution of, 142–143
        cooperative behavior, 143
        group selection theory, 142
        iterated interactions, 143
        kin selection, 141–142
        prosocial behavior, 141
        simulated conspecific, 144–145
        temptation payoff, 143
    "punishment" strategies, 147
    songbird species, 145–146
    tit for tat, 146
Garbage dump hypothesis, 13, 15
Generalized anxiety disorder, 25
Genetic disorder, 24–25
German shepherds, 160, 179–180, 191–192
Goldendoodle, 11
Golden retrievers, 131–132, 162–163
Grandin, Temple, 43–44, 66–67, 103–104
Grass eating behavior, 57–58
Graybeard, David, 97–98
Gray matter, 64–65
Gray wolf *(Canis lupus fuscus)*. *See* Wolf
Great Danes, 7–8
Greyhounds, 113–114, 162–163
Growl, 53, 59
Guilty, 33, 35–36, 53, 59–60

H

Hamilton, Alexander, 113–115
Hamilton's rule, 129–130
Hamilton, W.D., 142–143
Hamster *(Mesocricetus auratus)*, 162
Haplodiploidy, 113
Hawk-dove game, 150, 170
Hearing
  cats, 82, 87–88
  dogs, 43–44, 87–88
  humans, 87–88
Hemophilia, 27–28
Herding dogs, 23, 32
Historical hoax, 172
Hominini tribe, 124–125
Homunculus, 67
Honey hunting, 36
Hooded warblers *(Wilsonia citrina)*, 145–146, 145f
Horowitz, Alexandra, 100
Horse *(Equus caballus)*, 128–129, 136–137, 183, 188
Howling, 56
Human-canine coevolution, 99
Human emotions, 101
Humane Society of the United States (HSUS), 164
Humans
  brain, 65, 66f, 85
    average weighs, 65–66
    encephalization quotient, 65–66
    frontal lobe, 66–67
    Gage, Phineas, 63–64, 64f
    homunculus, 67
  canine relationship, xx, 13, 17
    anthropomorphism, 33–34
    Chauvet footprints, xviiif, xvii, xvii–xviii
    family members, xviii–xix, xxi
    flehmen response, 57
    food reward, 36
    food source, 12–13
    guilty/dog shaming, 33, 35–36
    high-risk behaviors, 36
    hunting partners, 16–17
    lupomorphizing, 33
    one/two-legged bicycle response, 57
    pet parents, xviii–xix, xxii–xxiii
    protectors, 16–17
    sensitive and critical periods of development, 73–75
    verbal and nonverbal communication, xix, 46–55, 58–61
  cats, relationship with, 17–19, 22
  chimpanzees, xix–xx
  encephalization quotient, 124–125
  facial microexpressions, 44, 46
  hearing, sense of, 87–88
  high-reward behavior, 36
  inbreeding, 26–28, 27–28f
  and nonhuman animals
    anthropomorphism, 33–34
    ethology, history of, 36–40
  pareidolia, 41
  sensory experience, 33
  sixth sense, 79
  smell, sense of
    olfactory receptors, 43, 85
    Proust phenomenon, 85–86
  taste, 81, 89–92
  umwelt, 41–42
  vision, 43–44, 80
  wolf, companionship with, 12, 15–16
Hunter-gatherer societies, 114–115
Hunting
  dogs, 16
  honey, 36
  mammoth, 36
Hyperkalemic periodic paralysis (HYPP), 28–29
Hypoallergenic, 11–12

I

Impressive, 28–29
Imprinting, 37–39
Inbreeding
  cheetahs, 25–26
  depression, 179–180
  dogs, 23, 25, 29–30
  domesticated animals, genetic disorders in, 28–29
  horses, 28–29
  humans, 26–28, 27–28f
  wild animals, genetic disorders in, 28–29
Inclusive fitness, 113
Influenza, 122
Information acquisition, 72
Innenwelt, xxiii, 40, 42
International Association for Animal Hospice and Palliative Care (IAAHPC), 187–188
Interspecific relationships, 34–35
Interspecific reproduction, 8–9
Intraspecies reproduction, 8–9
Isle Royale, 165–166
Italian greyhound, 5, 11
Iterated games, 144

J

Jack Russell Terriers, 162–163
Jacobson's organ, 43

K
King Tutankhamun, 27–28, 28f
K9s dogs, 157–158

L
Labradoodle, 11
Labrador, 131–132, 162–164, 189
Learning, dogs
   behaviorism, 72
   behaviorist learning theory, 72
   cognitivism, 72
   constructivism, 72
   early communicative skills, 72–73
   negative reinforcement and extinction, 71–72
   Pavlovian conditioning, 72
   positive punishment, 71
   salivation response, 72
Licking behaviors, 51–52, 56
Limbic brain, 66
Limburger cheese, 91
Lioness, 34–35, 35f

M
Mammoth hunting, 36
Manhattan Project, 139
Maternal DNA (mtDNA), 14–15
MBTI. *See* Myers-Briggs Type Indicator (MBTI)
McConnell, Patricia, 167
Measles, mumps, and rubella (MMR) vaccine, 164–165
Mech, David, 165–166
Mendel, Gregor, 127–128
Mexican Hairless Dogs, 113–114
Microemotions, 46
Microexpressions, 44, 46
Micromomentary expressions, 46
Millan, Cesar, 167
Milne, Emma Goodman, 30
Mimus melanotis. See Mockingbirds
Mind, 63–64
Mirror neuron system (MNS), 105–106
Mixed-breed dogs, 11–12
Mockingbirds, 1–3, 4f
Moths, 41f
   industrialized melanism with, 6, 7f
   umwelt, 41
Mouse *(Mus musculus)*, 162
Myers-Briggs Type Indicator (MBTI), 130–131, 131f, 134–135

N
Nash, John Forbes Jr., 140, 140f
Negative reinforcement, 71–72
Neotenization, 11
Neotenous facial expressions, 45
Neurological disorder, 105
Neurons, 64–66
Neurotransmitters, human aggression, 162, 162f
Noniterated game, 143–144, 184–185
Nonvocal mouth communication, 51–52

O
Obsessive-compulsive disorder (OCD), 103–104
Olfactory memory, 85–86
Olfactory receptors, 43, 80–81, 85
Operant conditioning, 39
OR6A2, 80–81
Orca
   encephalization quotient, 124–125
The Origin of Species, 1–2, 96–97, 113, 175
OR4N5, 80–81

P
Paedomorphism, 45
Pareidolia, 41
Parker, Geoff, 170
Parrots, 101
Parvovirus, 122
Pavlovian conditioning, 72, 109
Pavlov, Ivan, 109
Peacock, 6
Pekingese, 32
Pentadactylism, 42
Peppered moth, 6, 7f
Pet hypothesis, 13, 15
Physiological emotions, 93–94, 98–99
Pigeons, 3–4
Piltdown man *(Eoanthropus dawsoni)*, 171–173
Pit bulls, 163–164
Pit Bull Terriers, 162–163
Playback studies, 47
Play behaviors, 53–54
"Play-bow" position, 52–54, 52f
Pleiotropy, 127–128
Pomeranian, 163–164
Poodles, 131–132, 163–164
Positive punishment, 71, 77
Positive-reinforcement training, 77–78
Prairie vole *(Microtus ochrogaster)*, 162
Predatory drift, xxii
Primary emotions, 93–94
Principle of reinforcement, 39
Profound depression, 102
Proust phenomenon, 85–86
Psychology, 39

Pug, 5
Puggle, 11
Punishment, 71, 77–78
"Puppy-dog" eyes, 44–45, 49–50

R
Raccoon dogs, 7–8
Rafinesque, Constantine Samuel, 172
Rat *(Rattus norvegicus)*, 162
    DR2 cell activity, 110
    risk-averse tendencies, 110
Rattlesnake, 41
Reciprocal altruism, 111, 113–114, 149, 151
Red fox *(Vulpes fulva macroura)*, 173–175
Referential pointing, 154
Referential vocal communication, 47
Reinforcement, principle of, 39
Reptilian brain, 66
"Resource guarding," 101
Resource holding potential (RHP), 170
Risk-taking behaviors
    anticipatory salivation, 109
    brain's reward system, 110
    cost-benefit analysis, 111–112
    decision-making, 112
    evolutionary strategies, 111
    gambling, 109–111
    Hamilton's rule, 113–115
    intermittent reinforcement, 109–110
    kin selection, 111
    nonhuman animal behavior, 112
    pets, life of
        behaviors, 115–116, 118
        costs and benefits, 116–118
        distance, 118–119
        domestication, 115–116
        familiar dogs, 116–117
        jobs, 115–116
        motivation, 116
        overstimulation, 119
        positive and negative learning, 117–118
        punishment techniques, 117
        reciprocal altruism, 116
        social interactions, 118
        unfamiliar dogs, 116–117
    proximate and ultimate explanations, 111
    reciprocal altruism, 111
    risk-prone members, 114–115
    sister act, 112–113
    survival mechanism, 109
    taste aversion, 109
    time-consuming behaviors, 112
"Rock-paper-scissors," 139

S
Saint Bernards, 191–192
Salivation response, 72
Salukis, 191–192
Scale of nature, 37–39, 38f
Scramble competition, 156
Secondary emotions, 93–94
Self-calming behaviors, 59
Self-injurious behaviors, 104
Sensory perception
    anatomy, role in, 81–82
    hearing
        cats, 82, 87–88
        dogs, 43–44, 87–88
        humans, 87–88
    smell (*see* Smell)
    supernormal stimuli, 82
    taste, 80–81, 89–92
    touch, 88–89
    vision (*see* Vision)
Separation anxiety, 105
Sexual selection, 6
Siberian huskies, 179–180, 182
Side-botched lizard *(Uta stansburiana)*, 149–150, 149f
Sight, 83–85
Silver foxes, xx, 10–11
Smallpox, 122
Smell
    cats, 82, 85
    dogs, 42–44, 59–60, 86, 91–92
        accuracy, 81
        diseases, detection of, 86–87
        drugs and explosives, 42–43
        human-related scents, 85
        long snouts, 87
        olfactory memory, 85–86
        olfactory receptors, 43, 85
        pain and fear, detection of, 86–87
        sickness, detection of, 86–87
        three-dimensional, 42
        vomeronasal organ, 43
    humans, 43–44
        olfactory receptors, 85
        Proust phenomenon, 85–86
    olfactory receptor genes, 80–81
Smithsonian Institute of Natural History, 173, 174f
Social anxiety, 75
Sorokina, Nina, 126–127, 161
Sound, 43–44, 87–88
South American mockingbirds, 1–2
Springer rage syndrome, 24–25
Stegomastodon, 171

Stereotypies, 103–104
Stickleback fish *(Gasterosteus aculeatus)*, 144–145, 144f
Submissive position, 51–52
Supernormal stimuli, 82
Swift fox *(Vulpes velox)*, 173–175
Synapses, 64–65
Systema Naturae, 176–177

T

Tail chasing, 56
Tail movement, 49–50
Tamed apes, 126–127
Tamed fox, 126, 126f
TAS2R1, 80–81
Taste, 80–81, 89–92
Taste aversion, 109
Tennessee Cumberland Plateau, 175
Terriers, 113–114
Tetrachromatic vision, 83
Therapy dogs, 133, 189
The Third Chimpanzee, 124–125
Tick, 41, 41f
Touch, 88–89
Training, 32, 77–78
Trichromatic color vision, 83
Turnspit dogs, 181

U

Umwelt, xxiii, 40–42, 81, 84

V

Vampire bats *(Desmodus rotundus)*, 142
Vibrissae, 88–89
Vision
  color vision *(see* Color vision*)*
  dogs
    blind corridor, 84
    color blind, 83
    dichromatic vision, 83
    dog park designing, 84
    low-resolution vision, 43
    red-green differences, 83–84
    superior vision, 43
  humans, 43–44, 80
Visual gestures, 55
Vocal communication, 47, 58–59
Vomeronasal organ, 43
Von Neumann, John, 139
The Voyage of the Beagle, 2–3

W

Wakefield, Andrew, 164–165
Walam Olum/"red record," 172
Whippets, 162–163
White matter, 64–65
Wild children, 176
Wolf
  "alpha" concept, 166
  brain, 65–66
  cooperative behaviors, 156–157
  developmental milestones, 69
  divergent socioecology of, 156–157
  domesticated dogs, descent of, xix–xx, 7–8, 14–17
  dominance hierarchy, 156
  garbage dump hypothesis, 13
  hierarchies, 166
  historical images, 15, 16f
  humans, companionship with, 12, 15–16
  hybrids, behavioral issues, 9–11
  interspecific reproduction, 8–9
  Isle Royale, 165–166
  sanctioned hunts of, 15–16
  starch metabolizing genes, 13
Wolf Haven International, 177–178
The Wolf: The Ecology and Behavior of an Endangered Species, 165–166
Wolfhounds, 113–114

Y

Yawning, 51–52
Yellow fever, 122

9780128164983